BIOPHILIA

Soft, to your places, animals,
Your legendary duty calls.

THOMAS KINSELLA

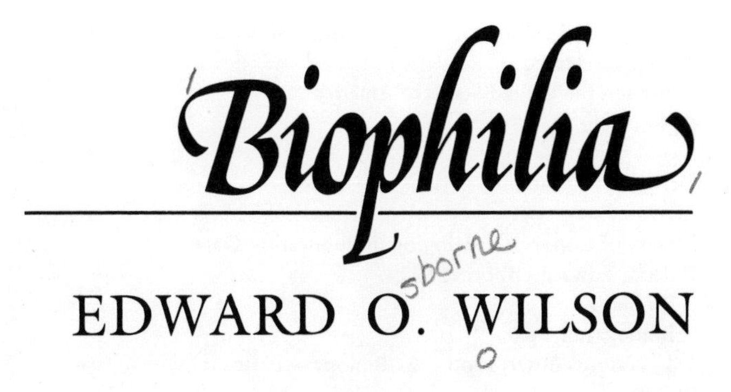

Biophilia

EDWARD O. WILSON

Harvard University Press
Cambridge, Massachusetts, and London, England

Printed in the United States of America
Eighth printing, 1996

Designed by Marianne Perlak

Library of Congress Cataloging in Publication Data
Wilson, Edward Osborne, 1929–
  Biophilia.
  Bibliography: p.
  1. Nature conservation. 2. Biology—Philosophy.
I. Title.
QH75.W534 1984 333.95′16 84-9052
ISBN 0-674-07441-6 (cloth)
ISBN 0-674-07442-4 (paper)

Contents

BIOPHILIA

Prologue

ON MARCH 12, 1961, I stood in the Arawak village of Bernhardsdorp and looked south across the white-sand coastal forest of Surinam. For reasons that were to take me twenty years to understand, that moment was fixed with uncommon urgency in my memory. The emotions I felt were to grow more poignant at each remembrance, and in the end they changed into rational conjectures about matters that had only a distant bearing on the original event.

The object of the reflection can be summarized by a single word, biophilia, which I will be so bold as to define as the innate tendency to focus on life and lifelike processes. Let me explain it very briefly here and then develop the larger theme as I go along.

From infancy we concentrate happily on ourselves and other organisms. We learn to distinguish life from the inanimate and move toward it like moths to a porch light. Novelty and diversity are particularly esteemed; the mere mention of the word *extraterrestrial* evokes reveries about still unexplored life, displacing the old and once potent *exotic* that drew earlier generations to remote islands and jungled interiors. That much is immediately clear, but a great deal more needs to be added. I will make the case that to explore and affiliate with life is a deep and complicated process in mental development. To an extent still undervalued in philosophy and religion, our existence depends on this propensity, our spirit is woven from it, hope rises on its currents.

There is more. Modern biology has produced a genuinely new way of looking at the world that is incidentally congenial to the inner direction of biophilia. In other words, instinct is in this rare instance aligned with reason. The conclusion I draw is optimistic: to the degree that we come to understand other organisms, we will place a greater value on them, and on ourselves.

Bernhardsdorp

AT BERNHARDSDORP on an otherwise ordinary tropical morning, the sunlight bore down harshly, the air was still and humid, and life appeared withdrawn and waiting. A single thunderhead lay on the horizon, its immense anvil shape diminished by distance, an intimation of the rainy season still two or three weeks away. A footpath tunneled through the trees and lianas, pointing toward the Saramacca River and far beyond, to the Orinoco and Amazon basins. The woodland around the village struggled up from the crystalline sands of the Zanderij formation. It was a miniature archipelago of glades and creekside forest enclosed by savanna—grassland with scattered trees and high bushes. To the south it expanded to become a continuous lacework fragmenting the savanna and transforming it in turn into an archipelago. Then, as if conjured upward by some unseen force, the woodland rose by stages into the triple-canopied rain forest, the principal habitat of South America's awesome ecological heartland.

In the village a woman walked slowly around an iron cooking pot, stirring the fire beneath with a soot-blackened machete. Plump and barefoot, about thirty years old, she wore two long pigtails and a new cotton dress in a rose floral print. From politeness, or perhaps just shyness, she gave no outward sign of recognition. I was an apparition, out of place and irrelevant, about to pass on down the footpath and out of her circle of required attention. At her feet a small child traced meanders in the dirt with a stick. The village around them was a cluster of no more than ten one-room dwellings. The walls were made of palm leaves woven into a herring-bone pattern in which dark bolts zigzagged upward and to the onlooker's right across flesh-colored squares. The design was the sole indigenous artifact on display. Bernhardsdorp was too close to Paramaribo, Surinam's capital, with its flood of cheap manufactured products to keep the look of a real Arawak village. In culture as in name, it had yielded to the colonial Dutch.

A tame peccary watched me with beady concentration from beneath the shadowed eaves of a house. With my own, taxonomist's eye I registered the defining traits of the collared species, *Dicotyles tajacu:* head too large for the piglike body, fur coarse and brindled, neck circled by a pale thin stripe, snout tapered, ears erect, tail reduced to a nub. Poised on stiff little dancer's legs, the young male seemed perpetually fierce and ready to charge yet frozen in place, like the metal boar on an ancient Gallic standard.

A note: Pigs, and presumably their close relatives the peccaries, are among the most intelligent of animals. Some biologists believe them to be brighter than dogs, roughly the rivals of elephants and porpoises. They form herds of ten to twenty members, restlessly patrolling territories of about a square mile. In certain ways they behave more like wolves and dogs than social ungulates. They recognize one another as individuals, sleep with their fur touching, and bark back and forth when on the move. The adults are organized into dominance orders in which the females are ascendant over males, the reverse of the usual mammalian arrangement. They attack in groups if cornered, their scapular fur bristling outward like porcupine quills, and can slash to the bone with sharp canine teeth. Yet individuals are easily tamed if captured as infants and their repertory stunted by the impoverishing constraints of human care.

So I felt uneasy—perhaps the word is embarrassed—in the presence of a captive individual. This young adult was a perfect anatomical specimen with only the rudiments of social behavior. But he was much more: a powerful presence, programed at birth to respond through learning steps in exactly the collared-peccary way and no other to the immemorial environment from which he had been stolen, now a mute speaker trapped inside the unnatural clearing, like a messenger to me from an unexplored world.

I stayed in the village only a few minutes. I had come to study ants and other social insects living in Surinam. No trivial task: over a hundred species of ants and termites are found within a square mile of average South American tropical forest. When all the animals in a randomly selected patch of woodland are collected together and weighed, from tapirs and parrots down to the smallest insects and roundworms, one third of the weight is found to consist of ants and termites. If you close your eyes and lay your hand on a tree trunk almost anywhere in the tropics until you feel something touch it, more times than not the crawler will be an ant. Kick open a rotting log and termites pour out. Drop a crumb of bread on the ground and within minutes ants of one kind or another drag it down a nest hole. Foraging ants are the chief predators of insects and other small animals in the tropical forest, and termites are the key animal decomposers of wood. Between them they form the conduit for a large part of the energy flowing through the forest. Sunlight to leaf to caterpillar to ant to anteater to jaguar to maggot to humus to termite to dissipated heat: such are the links that compose the great energy network around Surinam's villages.

I carried the standard equipment of a field biologist: camera; canvas satchel containing forceps, trowel, ax, mosquito repellent, jars, vials of alcohol, and notebook; a twenty-power hand lens swinging with a reassuring tug around the neck; partly fogged eyeglasses sliding down the nose and khaki shirt plastered to the back with sweat. My attention was on the forest; it has been there all my life. I can work up some appreciation for the travel stories of Paul Theroux and other urbanophile authors who treat human settlements as virtually the whole world and the intervening natural habitats as troublesome barriers. But everywhere I have gone—South America, Australia, New Guinea, Asia—I have thought exactly the opposite. Jungles and grasslands are the logical destinations, and towns and farmland the labyrinths that people have imposed between them sometime in the past. I cherish the green enclaves accidentally left behind.

Once on a tour of Old Jerusalem, standing near the elevated site of Solomon's Throne, I looked down across the Jericho Road to the dark olive trees of Gethsemane and wondered which native Palestinian plants and animals might still be found in the shade underneath. Thinking of "Go to the ant, thou sluggard; consider her ways," I knelt on the cobblestones to watch harvester ants carry seeds down holes to their subterranean granaries, the same food-gathering activity that had impressed the Old Testament writer, and possibly the same species at the very same place. As I walked with my host back past the Temple Mount toward the Muslim Quarter, I made inner calculations of the number of ant species found within the city walls. There was a perfect logic to such eccentricity: the million-year history of Jerusalem is at least as compelling as its past three thousand years.

AT BERNHARDSDORP I imagined richness and order as an intensity of light. The woman, child, and peccary turned into incandescent points. Around them the village became a black disk, relatively devoid of life, its artifacts adding next to nothing. The woodland beyond was a luminous bank, sparked here and there by the moving lights of birds, mammals, and larger insects.

I walked into the forest, struck as always by the coolness of the shade beneath tropical vegetation, and continued until I came to a small glade that opened onto the sandy path. I narrowed the world down to the span of a few meters. Again I tried to compose the mental set—call it the naturalist's trance, the hunter's trance—by which biologists locate more elusive organisms. I imagined that this place and all its treasures were mine alone and might be so forever in memory—if the bulldozer came.

In a twist my mind came free and I was aware of the hard workings of the natural world beyond the periphery of ordinary attention, where passions lose their meaning and history is in another dimension, without people, and great events pass without record or judgment. I was a transient of no consequence in this familiar yet deeply alien world that I had come to love. The uncounted products of evolution were gathered there for purposes having nothing to do with me; their long Cenozoic history was enciphered into a genetic code I could not understand. The effect was strangely calming. Breathing and heartbeat diminished, concentration intensified. It seemed to me that something extraordinary in the forest was very close to where I stood, moving to the surface and discovery.

I focused on a few centimeters of ground and vegetation. I willed animals to materialize, and they came erratically into view. Metallic-blue mosquitoes floated down from the canopy in search of a bare patch of skin, cockroaches with variegated wings perched butterfly-like on sunlit leaves, black carpenter ants sheathed in recumbent golden hairs filed in haste through moss on a rotting log. I turned my head slightly and all of them vanished. Together they composed only an infinitesimal fraction of the life actually present. The woods were a biological maelstrom of which only the surface could be scanned by the naked eye. Within my circle of vision, millions of unseen organisms died each second. Their destruction was swift and silent; no bodies thrashed about, no blood leaked into the ground. The microscopic bodies were broken apart in clean biochemical chops by predators and scavengers, then assimilated to create millions of new organisms, each second.

Ecologists speak of "chaotic regimes" that rise from orderly processes and give rise to others in turn during the passage of life from lower to higher levels of organization.

The forest was a tangled bank tumbling down to the grassland's border. Inside it was a living sea through which I moved like a diver groping across a littered floor. But I knew that all around me bits and pieces, the individual organisms and their populations, were working with extreme precision. A few of the species were locked together in forms of symbiosis so intricate that to pull out one would bring others spiraling to extinction. Such is the consequence of adaptation by coevolution, the reciprocal genetic change of species that interact with each other through many life cycles. Eliminate just one kind of tree out of hundreds in such a forest, and some of its pollinators, leafeaters, and woodborers will disappear with it, then various of their parasites and key predators, and perhaps a species of bat or bird that depends on its fruit—and when will the reverberations end? Perhaps not until a large part of the diversity of the forest collapses like an arch crumbling as the keystone is pulled away. More likely the effects will remain local, ending with a minor shift in the overall pattern of abundance among the numerous surviving species. In either case the effects are beyond the power of present-day ecologists to predict. It is enough to work on the assumption that all of the details matter in the end, in some unknown but vital way.

After the sun's energy is captured by the green plants, it flows through chains of organisms dendritically, like blood spreading from the arteries into networks of microscopic capillaries. It is in such capillaries, in the life cycles of thousands of individual species, that life's important work is done. Thus nothing in the whole system makes sense until the natural history of the constituent species becomes known. The study of every kind of organism matters, everywhere in the world. That conviction leads the field biologist to places like Surinam and the outer limits of evolution, of which this case is exemplary:

> The three-toed sloth feeds on leaves high in the canopy of the lowland forests through large portions of South and

Central America. Within its fur live tiny moths, the species *Cryptoses choloepi,* found nowhere else on Earth. When a sloth descends to the forest floor to defecate (once a week), female moths leave the fur briefly to deposit their eggs on the fresh dung. The emerging caterpillars build nests of silk and start to feed. Three weeks later they complete their development by turning into adult moths, and then fly up into the canopy in search of sloths. By living directly on the bodies of the sloths, the adult *Cryptoses* assure their offspring first crack at the nutrient-rich excrement and a competitive advantage over the myriad of other coprophages.

At Bernhardsdorp the sun passed behind a small cloud and the woodland darkened. For a moment all that marvelous environment was leveled and subdued. The sun came out again and shattered the vegetative surfaces into light-based niches. They included intensely lighted leaf tops and the tops of miniature canyons cutting vertically through tree bark to create shadowed depths two or three centimeters below. The light filtered down from above as it does in the sea, giving out permanently in the lowermost recesses of buttressed tree trunks and penetralia of the soil and rotting leaves. As the light's intensity rose and fell with the transit of the sun, silverfish, beetles, spiders, bark lice, and other creatures were summoned from their sanctuaries and retreated back in alternation. They responded according to receptor thresholds built into their eyes and brains, filtering devices that differ from one kind of animal to another. By such inborn controls the species imposed a kind of prudent self-discipline. They unconsciously halted their population growth before squeezing out competitors, and others did the same. No altruism was needed to achieve this balance, only specialization. Coexistence was an incidental by-product of the Darwinian advantage that accrued from the avoidance of competition. During the long span of evolution the species divided the environment among themselves, so that now each tenuously preempted certain of the capillaries of energy flow. Through repeated genetic changes they sidestepped competitors and built elaborate defenses against the host of predator species that relentlessly tracked them through matching genetic countermoves. The result was a splendid array of specialists, including moths that live in the fur of three-toed sloths.

NOW TO THE very heart of wonder. Because species diversity was created prior to humanity, and because we evolved within it, we have never fathomed its limits. As a consequence, the living world is the natural domain of the most restless and paradoxical part of the human spirit. Our sense of wonder grows exponentially: the greater the knowledge, the deeper the mystery and the more we seek knowledge to create new mystery. This catalytic reaction, seemingly an inborn human trait, draws us perpetually forward in a search for new places and new life. Nature is to be mastered, but (we hope) never completely. A quiet passion burns, not for total control but for the sensation of constant advance.

At Bernhardsdorp I tried to convert this notion into a form that would satisfy a private need. My mind maneuvered through an unending world suited to the naturalist. I looked in reverie down the path through the savanna woodland and imagined walking to the Saramacca River and beyond, over the horizon, into a timeless reconnaissance through virgin forests to the land of magical names, Yékwana, Jívaro, Sirionó, Tapirapé, Siona-Secoya, Yumana, back and forth, never to run out of fresh jungle paths and glades.

The same archetypal image has been shared in variations by others, and most vividly during the colonization of the New World. It comes through clearly as the receding valleys and frontier trails of nineteenth-century landscape art in the paintings of Albert Bierstadt, Frederick Edwin Church, Thomas Cole, and their contemporaries during the crossing of the American West and the innermost reaches of South America.

In Bierstadt's *Sunset in Yosemite Valley* (1868), you look down a slope that eases onto the level valley floor, where a river flows quietly away through waist-high grass, thickets, and scattered trees. The sun is near the horizon. Its dying light, washing the surface in reddish gold, has begun to yield to blackish green shadows along the near side of the valley. A cloud bank has lowered to just beneath the tops of the sheer rock walls. More protective than threatening, it has transformed the valley into a tunnel opening out through the far end into a sweep of land. The world beyond is obscured by the blaze of the setting sun into which we are forced to gaze in order to see that far. The valley, empty of people, is safe: no fences, no paths, no owners. In a few minutes we could walk to the river, make camp, and afterward explore away from the banks at leisure. The ground in sight is human-sized, measured literally by foot strides and strange new plants and animals large enough to be studied at twenty paces. The dreamlike quality of the painting rolls time forward: what might the morning bring? History is still young, and human imagination has not yet been chained by precise geographic knowledge. Whenever we wish, we can strike out through the valley to the unknown terrain beyond, to a borderland of still conceivable prodigies—bottomless vales and boundless floods, in Edgar Allan Poe's excited imagery, "and chasms, and caves and Titan woods with forms that no man can discover." The American frontier called up the old emotions that had pulled human populations like a living sheet over the world during the ice ages. The still unfallen western world, as Melville wrote of the symbolizing White Steed in *Moby Dick,* "revived the glories of those primeval times when Adam walked majestic as a god."

Then a tragedy: this image is almost gone. Although perhaps as old as man, it has faded during our own lifetime. The wildernesses of the world have shriveled into timber leases and threatened nature reserves. Their parlous state presents us with a dilemma, which the historian Leo Marx has called the machine in the garden. The natural world is the refuge of the spirit, remote, static, richer even than human imagination. But we cannot exist in this paradise without the machine that tears it apart. We are killing the thing we love, our Eden, progenitrix, and sibyl. Human beings are not captive peccaries, natural creatures torn from a sylvan niche and imprisoned within a world of artifacts. The noble savage, a biological impossibility, never existed. The human relation to nature is vastly more subtle and ambivalent, probably for this reason. Over thousands of generations the mind evolved within a ripening culture, creating itself out of symbols and tools, and genetic advantage accrued from planned modifications of the environment. The unique operations of the brain are the result of natural selection operating through the filter of culture. They have suspended us between the two antipodal ideals of nature and machine, forest and city, the natural and the artifactual, relentlessly seeking, in the words of the geographer Yi-Fu Tuan, an equilibrium not of this world.

So at Bernhardsdorp my own thoughts were inconstant. They skipped south to the Saramacca and on deep into the Amazon basin, the least spoiled garden on Earth, and then swiftly back north to Paramaribo and New York, greatest of machines. The machine had taken me there, and if I ever seriously thought of confronting nature without the conveniences of civilization, reality soon regained my whole attention. The living sea is full of miniature horrors designed to reduce visiting biologists to their constituent amino acids in quick time. Arboviruses visit the careless intruder with a dismaying variety of chills and diarrhea. Breakbone fever swells the joints to agonizing tightness. Skin ulcers spread remorselessly outward from thorn scratches on the ankle. Triatoma assassin bugs suck blood from the sleeper's face during the night and leave behind the fatal microorganisms of Chagas' disease—surely history's most unfair exchange. Leishmaniasis, schistosomiasis, malignant tertian malaria, filariasis, echinococcosis, onchocerciasis, yellow fever, amoebic dysentery, bleeding bot-fly cysts . . . evolution has devised a hundred ways to macerate livers and turn blood into a parasite's broth. So the romantic voyager swallows chloraquin, gratefully accepts gamma globulin shots, sleeps under mosquito netting, and remembers to pull on rubber boots before wading in freshwater streams. He hopes that enough fuel was put into the Land Rover that morning, and he hurries back to camp in time for a hot meal at dusk.

The impossible dilemma caused no problem for ancestral men. For millions of years human beings simply went at nature with everything they had, scrounging food and fighting off predators across a known world of a few square miles. Life was short, fate terrifying, and reproduction an urgent priority: children, if freely conceived, just about replaced the family members who seemed to be dying all the time. The population flickered around equilibrium, and sometimes whole bands became extinct. Nature was something out there—nameless and limitless, a force to beat against, cajole, and exploit.

If the machine gave no quarter, it was also too weak to break the wilderness. But no matter: the ambiguity of the opposing ideals was a superb strategy for survival, just so long as the people who used it stayed sufficiently ignorant. It enhanced the genetic evolution of the brain and generated more and better culture. The world began to yield, first to the agriculturists and then to technicians, merchants, and circumnavigators. Humanity accelerated toward the machine antipode, heedless of the natural desire of the mind to keep the opposite as well. Now we are near the end. The inner voice murmurs *You went too far,* and disturbed the world, and gave away too much for your control of Nature. Perhaps Hobbes's definition is correct, and this will be the hell we earned for realizing truth too late. But I demur in all this. I suggest otherwise: the same knowledge that brought the dilemma to its climax contains the solution. Think of scooping up a handful of soil and leaf litter and spreading it out on a white ground cloth, in the manner of the field biologist, for close examination. This unprepossessing lump contains more order and richness of structure, and particularity of history, than the entire surfaces of all the other (lifeless) planets. It is a miniature wilderness that can take almost forever to explore.

Tease apart the adhesive grains with the aid of forceps, and you will expose the tangled rootlets of a flowering plant, curling around the rotting veins of humus, and perhaps some larger object such as the boat-shaped husk of a seed. Almost certainly among them will be a scattering of creatures that measure the world in millimeters and treat this soil sample as traversable: ants, spiders, springtails, armored oribatid mites, enchytraeid worms, millipedes. With the aid of a dissecting microscope, proceed on down the size scale to the roundworms, a world of scavengers and fanged predators feeding on them. In the hand-held microcosm all these creatures are still giants in a relative sense. The organisms of greatest diversity and numbers are invisible or nearly so. When the soil-and-litter clump is progressively magnified, first with a compound light microscope and then with scanning electron micrographs, specks of dead leaf expand into mountain ranges and canyons, soil particles become heaps of boulders. A droplet of moisture trapped between root hairs grows into an underground lake, surrounded by a three-dimensional swamp of moistened humus. The niches are defined by both topography and nuances in chemistry, light, and temperature shifting across fractions of a millimeter. Organisms now come into view for which the soil sample is a complete world. In certain places are found the fungi: cellular slime molds, the one-celled chitin-producing chytrids, minute gonapodyaceous and oomycete soil specialists, Kickxellales, Eccrinales, Endomycetales, and Zoopagales. Contrary to their popular reputation, the fungi are not formless blobs, but exquisitely structured organisms with elaborate life cycles. The following is a recently discovered extreme specialization, the example of the sloth moth repeated on a microscopic scale:

> In water films and droplets, attack cells of an oomycete, *Haptoglossa mirabilis,* await the approach of small, fat

wormlike animals the biologists call rotifers. Each cell is shaped like a gun; its anterior end is elongated to form a barrel, which is hollowed out to form a bore. At the base of the bore is a complicated explosive device. When a rotifer swims close, the attack cell detects its characteristic odor and fires a projectile of infective tissue through the barrel and into its body. The fungal cells proliferate through the victim's tissues and then metamorphose into a cylindrical fruiting body, from which exit tubes sprout. Next tiny spores separate themselves inside the fruiting body, swim out the exit tubes with the aid of whip-shaped hairs, and settle down to form new attack cells. They await more rotifers, prepared to trigger the soundless explosion that will commence a new life cycle.

Still smaller than the parasitic fungi are the bacteria, including colony-forming polyangiaceous species, specialized predators that consume other bacteria. All around them live rich mixtures of rods, cocci, coryneforms, and slime azotobacteria. Together these microorganisms metabolize the entire spectrum of live and dead tissue. At the moment of discovery some are actively growing and fissioning, while others lie dormant in wait for the right combination of nutrient chemicals. Each species is kept at equilibrium by the harshness of the environment. Any one, if allowed to expand without restriction for a few weeks, would multiply exponentially, faster and faster, until it weighed more than the entire Earth. But in reality the individual organism simply dissolves and assimilates whatever appropriate fragments of plants and animals come to rest near it. If the newfound meal is large enough, it may succeed in growing and reproducing briefly before receding back into the more normal state of physiological quiescence.

Biologists, to put the matter as directly as possible, have begun a second reconnaissance into the land of magical names. In exploring life they have commenced a pioneering adventure with no imaginable end. The abundance of organisms increases downward by level, like layers in a pyramid.

The handful of soil and litter is home for hundreds of insects, nematode worms, and other larger creatures, about a million fungi, and ten billion bacteria. Each of the species of these organisms has a distinct life cycle fitted, as in the case of the predatory fungus, to the portion of the microenvironment in which it thrives and reproduces. The particularity is due to the fact that it is programed by an exact sequence of nucleotides, the ultimate molecular units of the genes.

The amount of information in the sequence can be measured in bits. One bit is the information required to determine which of two equally likely alternatives is chosen, such as heads or tails in a coin toss. English words average two bits per letter. A single bacterium possesses about ten million bits of genetic information, a fungus one billion, and an insect from one to ten billion bits according to species. If the information in just one insect—say an ant or beetle—were to be translated into a code of English words and printed in letters of standard size, the string would stretch over a thousand miles. Our lump of earth contains information that would just about fill all fifteen editions of the *Encyclopaedia Britannica.*

To see what such molecular information can do, consider a column of ants running across the floor of a South American forest. Riding on the backs of some of the foragers are minute workers of the kind usually confined to duties within the underground nursery chambers. The full significance of hitchhiking is problematic, but at the very least the act helps to protect the colony against parasites. Tiny flies, members of the family Phoridae, hover above the running foragers. From time to time a fly dives down to thrust an egg into the neck of one of them. Later the egg hatches into a maggot that burrows deeper into the ant's body. The maggot grows rapidly, transforms into a pupa, and eventually erupts through the cuticle as an adult fly to restart the life cycle. The divebombers find the runners easy targets when they are burdened with a fragment of food. But when one also carries a hitchhiker, the smaller ant is able to chase the intruder away with its jaws and legs. It serves as a living fly whisk.

The brain of the fly or of the fly-whisk ant, when dissected out and placed in a drop of saline solution on a glass slide, resembles a grain of sugar. Although barely visible to the naked eye, it is a complete command center that choreographs the insect's movements through its entire adult cycle. It signals the precise hour for the adult to emerge from the pupal case; it processes the flood of signals transduced to it by the outer sensors; and it directs the performance of about twenty behavioral acts through nerves in the legs, antennae, and mandibles. The fly and the ant are hardwired in a manner unique to their respective species and hence radically different from each other, so that predator is implacably directed against prey, flier against runner, solitaire against colony member.

With advanced techniques it has been possible to begin mapping insect nervous systems in sufficient detail to draw the equivalent of wiring diagrams. Each brain consists of somewhere between a hundred thousand and a million nerve cells, most of which send branches to a thousand or more of their neighbors. Depending on their location, individual cells appear to be programed to assume a particular shape and to transmit messages only when stimulated by coded discharges from neighbor units that feed into them. In the course of evolution, the entire system has been miniaturized to an extreme. The fatty sheaths surrounding the axon shafts of the kind found in larger animals have been largely stripped away, while the cell bodies are squeezed off to one side of the multitudinous nerve connections. Biologists understand in very general terms how the insect brain might work as a complete on-board computer, but they are a long way from explaining or duplicating such a device in any detail.

The great German zoologist Karl von Frisch once said of his favorite organism that the honeybee is like a magic well: the more you draw from it, the more there is to draw. But science is in no other way mystical. Its social structure is such that anyone can follow most enterprises composing it, as observer if not as participant, and soon you find yourself on the boundaries of knowledge.

You start with the known: in the case of the honeybee, where it nests, its foraging expeditions, and its life cycle. Most remarkable at this level is the waggle dance discovered by von Frisch, the tail-wagging movement performed inside the hive to inform nestmates of the location of newly discovered flower patches and nest sites. The dance is the closest approach known in the animal kingdom to a true symbolic language. Over and over again the bee traces a short line on the vertical surface of the comb, while sister workers crowd in close behind. To return to the start of the line, the bee loops back first to the left and then to the right and so produces a figure-eight. The center line contains the message. Its length symbolically represents the distance from the hive to the goal, and its angle away from a line drawn straight up on the comb, in other words away from twelve o'clock, represents the angle to follow right or left of the sun when leaving the hive. If the bee dances straight up the surface of the comb, she is telling the others to fly toward the sun. If she dances ten degrees to the right, she causes them to go ten degrees right of the sun. Using such directions alone, the members of the hive are able to harvest nectar and pollen from flowers three miles or more from the hive.

The revelation of the waggle-dance code has pointed the way to deeper levels of biological investigation, and a hundred new questions. How does the bee judge gravity while on the darkened comb? What does it use for a guide when the sun goes behind a cloud? Is the waggle dance inherited or must it be learned? The answers create new concepts that generate still more mysteries. To pursue them (and we are now certainly at the frontier) investigators must literally enter the bee itself, exploring its nervous system, the interplay of its hormones and behavior, the processing of chemical cues by its nervous system. At the level of cell and tissue, the interior of the body will prove more technically challenging than the external workings of the colony first glimpsed. We are in the presence of a biological machine so complicated that to understand just one part of it—wings, heart, ovary, brain—can consume many lifetimes of original investigation.

And if that venture were somehow to be finished, it will merely lead on down into the essence of the machine, to the interior of cells and the giant molecules that compose their distinctive parts. Questions about process and meaning then take center stage. What commits an embryonic cell to become part of the brain instead of a respiratory unit? Why does the mother's blood invest yolk in the growing egg? Where are the genes that control behavior? Even in the unlikely event that all this microscopic domain is successfully mapped, the quest still lies mostly ahead. The honeybee, *Apis mellifera,* is the product of a particular history. Through fossil remains in rock and amber, we know that its lineage goes back at least 50 million years. Its contemporary genes were assembled by an astronomical number of events that sorted and recombined the constituent nucleotides. The species evolved as the outcome of hourly contacts with thousands of other kinds of plants and animals along the way. Its range expanded and contracted across Africa and Eurasia in a manner reminiscent of the fortunes of a human tribe. Virtually all this history remains unknown. It can be pursued to any length by those who take a special interest in *Apis mellifera* and seek what Charles Butler called its "most sweet and sov'raigne fruits" when he launched the modern scientific study of the honeybee in 1609.

Every species is a magic well. Biologists have until recently been satisfied with the estimate that there are between three and ten million of them on Earth. Now many believe that ten million is too low. The upward revision has been encouraged by the increasingly successful penetration of the last great unexplored environment of the planet, the canopy of the tropical rain forest, and the discovery of an unexpected number of new species living there. This layer is a sea of branches, leaves, and flowers crisscrossed by lianas and suspended about one hundred feet above the ground. It is one of the easiest habitats to locate—from a distance at least—but next to the deep sea the most difficult to reach. The tree trunks are thick, arrow-straight, and either slippery smooth or covered with sharp tubercles. Anyone negotiating them safely to the top must then contend with swarms of stinging ants and wasps. A few athletic and adventurous younger biologists have begun to overcome the difficulties by constructing special pulleys, rope catwalks, and observation platforms from which they can watch high arboreal animals in an undisturbed state. Others have found a way to sample the insects, spiders, and other arthropods with insecticides and quick-acting knockdown agents. They first shoot lines up into the canopy, then hoist the chemicals up in canisters and spray them out into the surrounding vegetation by remote control devices. The falling insects and other organisms are caught in sheets spread over the ground. The creatures discovered by these two methods have proved to be highly specialized in their food habits, the part of the tree in which they live, and the time of the year when they are active. So an unexpectedly large number of different kinds are able to coexist. Hundreds can fit comfortably together in a single tree top. On the basis of a preliminary statistical projection from these data, Terry L. Erwin, an entomologist at the National Museum of Natural History, has estimated that there may be thirty million species of insects in the world, most limited to the upper vegetation of the tropical forests.

Although such rough approximations of the diversity of life are not too difficult to make, the exact number of species is beyond reach because—incredibly—the majority have yet to be discovered and specimens placed in museums. Furthermore, among those already classified no more than a dozen have been studied as well as the honeybee. Even *Homo sapiens,* the focus of billions of dollars of research annually, remains a seemingly intractable mystery. All of man's troubles may well arise, as Vercors suggested in *You Shall Know Them,* from the fact that we do not know what we are and do not agree on what we want to become. This crucial inadequacy is not likely to be remedied until we have a better grasp of the diversity of the life that created and sustains us. So why hold back? It is a frontier literally at our fingertips, and the one for which our spirit appears to have been explicitly designed.

I WALKED on through the woodland at Bernhardsdorp to see what the day had to offer. In a decaying log I found a species of ant previously known only from the midnight zone of a cave in Trinidad. With the aid of my hand lens I identified it from its unique combination of teeth, spines, and body sculpture. A month before I had hiked across five miles of foothills in central Trinidad to find it in the original underground habitat. Now suddenly here it was again, nesting and foraging in the open. Scratch from the list what had been considered the only "true" cave ant in the world — possessed of workers pale yellow, nearly eyeless, and sluggish in movement. Scratch the scientific name *Spelaeomyrmex,* meaning literally cave ant, as a separate taxonomic entity. I knew that it would have to be classified elsewhere, into a larger and more conventional genus called *Erebomyrma,* ant of Hades. A small quick victory, to be reported later in a technical journal that specializes on such topics and is read by perhaps a dozen fellow myrmecologists. I turned to watch some huge-eyed ants with the formidable name *Gigantiops destructor.* When I gave one of the foraging workers a freshly killed termite, it ran off in a straight line across the forest floor. Thirty feet away it vanished into a small hollow tree branch that was partly covered by decaying leaves. Inside the central cavity I found a dozen workers and their mother queen — one of the first colonies of this unusual insect ever recorded. All in all, the excursion had been more productive than average. Like a prospector obsessed with ore samples, hoping for gold, I gathered a few more promising specimens in vials of ethyl alcohol and headed home, through the village and out onto the paved road leading north to Paramaribo.

Later I set the day in my memory with its parts preserved

for retrieval and closer inspection. Mundane events acquired the raiment of symbolism, and this is what I concluded from them: That the naturalist's journey has only begun and for all intents and purposes will go on forever. That it is possible to spend a lifetime in a magellanic voyage around the trunk of a single tree. That as the exploration is pressed, it will engage more of the things close to the human heart and spirit. And if this much is true, it seems possible that the naturalist's vision is only a specialized product of a biophilic instinct shared by all, that it can be elaborated to benefit more and more people. Humanity is exalted not because we are so far above other living creatures, but because knowing them well elevates the very concept of life.

The Superorganism

IN MARCH 1983 I returned to South America to begin a new program of study on tropical ants. I was interested in the way the communication systems and division of labor of these insects adapt them to their environment. My first stop was the field site of the Minimum Critical Size Project of the World Wildlife Fund, located in Amazonian forest sixty miles north of Manaus, Brazil. I was accompanied by Thomas Lovejoy, the young and vigorous vice-president for science of WWF-US, who had conceived the project in the late 1970s. We joined an assortment of researchers, students, and assistants who were working back and forth between Manaus and the site on weekly tours. Camaraderie came easily and was genuine; our shared values were implicit, forming a bond too strong to allow much discussion of why we had come together in this unlikely place. We talked only about organisms, in endless technical detail.

My hosts were not ordinary field biologists. They did not affect the verbal delicacy and critical reserve of typical academics encountered on leave from comfortable bases in Berkeley, Ann Arbor, and Cambridge. Their manner was self-confident and achievement-oriented, tough in a pleasant way, and they reminded me a bit of settlers I had met in Australia and New Guinea (and the Israeli biologist who pointed out the house where he had commanded a company during the 1967 war, as we returned from a field trip to the Dead Sea). Even though the World Wildlife Fund is operating on a slender budget, its Amazon Project is truly pioneering on a large scale. Planned to run into the next century, it is designed to answer one of the key questions of ecology and conservation practice: how extensive does a wildlife preserve have to be to sustain permanently most or all of the kinds of plants and animals protected within its boundaries?

We know that when a species loses part of its range, it is in greater danger of extinction. Expressed in loosely mathematical terms, the chance that a population of organisms will go extinct in a given year increases as its living space is cut back and its numbers are held at correspondingly lower numbers. All populations fluctuate in size to some extent, but those kept at a low maximum are more likely to zigzag all the way down to zero than those permitted to fluctuate at higher levels. For example, a population of ten grizzly bears living on one hundred square miles of land will probably disappear much earlier than a population of a thousand grizzly bears living on ten thousand square miles of similar land; the thousand could persist for centuries or, so far as ordinary human awareness is concerned, forever.

This simple fact of nature bears heavily on the design of nature reserves. When a piece of primeval forest is set aside and the surrounding forest cleared, it becomes an island in an agricultural sea. Like wave-lapped Puerto Rico or Bali, it has lost most of its connections with other natural land habitats from which new immigrations can occur. Over a period of years the number of plant and animal species will fall to a new and predictable level. Some impoverishment of diversity is inevitable even if men never put an ax to a single tree in the reserve. This natural decline presents biologists with a problem that is technically difficult and soluble only through the chancing of risky compromises. The reserve they recommend must be small enough to be economically reasonable. They cannot ask that an entire country be set aside. But they have the obligation to insist that the patch be made large enough to sustain the fauna and flora. It is their job to prove that a certain minimum area is required, to list as completely as possible which species will be saved in the reserve, and for approximately how long.

The tropical rain forest north of Manaus, like that in many other parts of the Amazon basin, is being clear-cut from the edge inward. It is being lifted up from the ground entire like a carpet rolled off a bare floor, leaving behind vast stretches of cattle range and cropland that need artificial fertilization to sustain even marginal productivity for more than two or three years. A rain forest in Brazil differs fundamentally from a deciduous woodland in Pennsylvania or Germany in the way its key resources are distributed. A much greater fraction of organic matter is bound up in the tissues of the standing trees, so that the leaf litter and humus are only a few inches deep. When the forest is felled and burned, the hard equatorial downpours quickly wash away the thin blanket of top soil.

Although I had this general information in advance, I was still shaken by the sight of newly cleared land around Manaus. The pans and hillocks of lateritic clay, littered with blackened tree stumps, bore the look of a freshly deserted battlefield. Spherical termite mounds sprouted from the fallen wood in an ill-fated population explosion, while vultures and swifts wheeled overhead in representation of the mostly vanished bird fauna. Bony white cattle, forlorn replacements of a magnificent heritage, clustered in small groups around the scattered watersheds. Near midday the heat of the sun bounced up from the bare patches of soil to hit with an almost tactile force. It was another world altogether from the shadowed tunnels of the nearby forest, and a constant reminder of what had happened: tens of thousands of species had been scraped away as by a giant hand and will not be seen in that place for generations, if ever. The action can be defended (with difficulty) on economic grounds, but it is like burning a Renaissance painting to cook dinner.

The Brazilian authorities have sanctioned the opening of the wilderness on the basis of a logical formula: the impoverished Northeast has people and no land, the Amazon has land and no people; join them together and build a nation. But they are also well aware of the problems of environmental degradation. More recently, influenced by biologists such as Warwick Kerr, Paulo Nogueira Neto, and Paulo Vanzolini, they have begun to forge a policy of preservation. Now by law, honored at least in principle, half of the forest must be left standing. Of equal importance, more than twenty Amazonian reserves and parks have been set aside in key areas where the greatest number of species of plants and animals are thought to exist. Most are over a thousand square miles in extent, the minimum area that, according to Princeton University's John Terborgh and other experts on the subject, is needed to hold the number of disappearing species to less than 1 percent of the initial complement over the next century. In other words, the formula is meant to ensure that 99 out of every 100 kinds of organisms will still be present in the year 2100. With reserves of this size there is hope even for the harpy eagle and the jaguar, of which single individuals need three square miles or more to survive, as well as the flourish of rare orchids, monkeys, river fish, and brilliant toucans and macaws that symbolize the admirable élan of Brazil itself.

But this leaves the smaller reserves elsewhere in Brazil and in the more densely populated countries of South and Central America. The American and Brazilian scientists at the Manaus project are addressing the problem in the following way. At the edge of the cutting, where the rain forest begins and continues virtually unbroken north to Venezuela, they have marked off a series of twenty plots ranging in size from one to a thousand hectares (a hectare is 100 meters on the side and equals 2.47 acres). In each plot they survey the most easily classified and monitored of the big organisms, including the trees, butterflies, birds, monkeys, and other large mammals. Then, with the cooperation of the landowners, they supervise the clearing of the surrounding land, leaving the plots behind as forest islands in a newly created agricultural sea. The surveys were begun in 1980 and will be continued for many years. Eventually the data should reveal how much faster the smaller island-reserves lose their species than larger ones, which kinds of animals and plants decline the most rapidly, why they become extinct, and, most crucially, the minimum area needed to hold on to the greater part of the diversity of life. No process being addressed by modern science is more complicated or, in my opinion, more important.

WE RODE in a World Wildlife Fund truck to a camp just inside the forest border at Fazenda Esteio, where the biologists were conducting one of the initial surveys. True to the philosophy of the sponsoring organization, the camp was a tiny clearing with a temporary shelter just large enough to hold a few hammocks, plus a cook's shed and fireplace, and nothing more. To my delight I found that I could roll out of my hammock in the morning, take twenty steps, and be in virgin rain forest. For five days I stayed in the woods except for meals and as little sleep as I needed to keep going.

I savored the cathedral feeling expressed by Darwin in 1832 when he first encountered tropical forest near Rio de Janeiro ("wonder, astonishment & sublime devotion, fill & elevate the mind"). And once again I could hold still for long intervals to study a few centimeters of tree trunk or ground, finding some new organism at each shift of focus. The intervals of total silence, often prolonged, became evidence of the intensity of the enveloping life. Several times a day I heard what may be the most distinctive sound of the primary tropical forest: a sharp *crack* like a rifle shot, followed by a whoosh and a solid thump. Somewhere a large tree, weakened by age and rot and top heavy from layers of vines, has chosen that moment to fall and end decades or centuries of life. The process is random and continuous, a sprinkling of events through the undisturbed portion of the forest. The broad trunk snaps or keels over to lever up the massive root system, the branches plow down through the canopies of neighboring trees at terrifying speed, and the whole thunders to the ground in a cloud of leaves, trailing lianas, and fluttering insects. There may be a hundred thousand trees within earshot of any place the hiker stands in the forest, so that the odds of hearing one coming down on a given day are high. But the chance of being close enough to be struck by any part of the tree is remote, comparable to that of stepping on a poisonous snake or coming round the bend of a trail one day to meet a mother jaguar with cubs. Still, the lifetime risk builds up cumulatively, like that in daily flights aboard single-engined airplanes, so that those who spend years in the forest count falling trees as an important source of danger.

Most of the time I worked with a restless energy to get ahead on several research projects I had in mind. I opened logs and twigs like presents on Christmas morning, entranced by the endless variety of insects and other small creatures that scuttled away to safety. None of these organisms was repulsive to me; each was beautiful, with a name and special meaning. It is the naturalist's privilege to choose almost any kind of plant or animal for examination and be able to commence productive work within a relatively short time. In the tropical forest, with thousands of mostly unknown species all around, the number of discoveries per investigator per day is probably greater than anywhere else in the world.

As if to dramatize the point, an insect I most wanted to find made its appearance soon after my arrival at Fazenda Esteio, with no effort of my own and literally at my feet. It was the leafcutter ant *(Atta cephalotes),* one of the most abundant and visually striking animals of the New World tropics. The *saúva,* as it is called locally, is a prime consumer of fresh vegetation, rivaled only by man, and a leading agricultural pest in Brazil. I had devoted years of research to the species in the laboratory but never studied it in the field. At dusk on the first day in camp, as the light failed to the point where we found it difficult to make out small objects on the ground, the first worker ants came scurrying purposefully out of the surrounding forest. They were brick red in color, about a quarter inch in length, and bristling with short, sharp spines.

Within minutes, several hundred had arrived and formed two irregular files that passed on either side of the hammock shelter. They ran in a nearly straight line across the clearing, their paired antennae scanning right and left, as though drawn by some directional beam from the other side. Within an hour, the trickle expanded to twin rivers of tens of thousands of ants running ten or more abreast. The columns could be traced easily with the aid of a flashlight. They came up from a huge earthen nest a hundred yards from the camp on a descending slope, crossed the clearing, and disappeared again into the forest. By climbing through tangled undergrowth we were able to locate one of their main targets, a tall tree bearing white flowers high in its crown. The ants streamed up the trunk, scissored out pieces of leaves and petals with their sharp-toothed mandibles, and headed home carrying the fragments over their heads like little parasols. Some floated the pieces to the ground, where most were picked up and carried away by newly arriving nestmates. At maximum activity, shortly before midnight, the trails were a tumult of ants bobbing and weaving past each other like miniature mechanical toys.

For many visitors to the forest, even experienced naturalists, the foraging expeditions are the whole of the matter, and individual leafcutter ants seem to be inconsequential ruddy specks on a pointless mission. But a closer look transforms them into beings of another order. If we magnify the scene to human scale, so that an ant's quarter-inch length grows into six feet, the forager runs along the trail for a distance of about ten miles at a velocity of 16 miles an hour. Each successive mile is covered in three minutes and forty-five seconds, about the current (human) world record. The forager picks up a burden of 750 pounds and speeds back toward the nest at 15 miles an hour—hence, four-minute miles. This marathon is repeated many times during the night and in many localities on through the day as well.

From research conducted jointly by biologists and chemists, it is known that the ants are guided by a secretion paid onto the soil through the sting, in the manner of ink being drawn out of a pen. The crucial molecule is methyl-4-methyl-pyrrole-2-carboxylate, which is composed of a tight ring of carbon and nitrogen atoms with short side chains made of carbon and oxygen. The pure substance has an innocuous odor, judged by various people to be faintly grassy, sulphurous, or fruitlike with a hint of naphtha (I'm not sure I can smell it at all). But whatever the impact on human beings, it is an ichor of extraordinary power for the ants. One milligram, a quantity that would just about cover the printed letters in this sentence, if dispensed with theoretical maximum efficiency, is enough to excite billions of workers into activity or to lead a short column of them three times around the world. The vast difference between us and them has nothing to do with the trail substance itself, which is a biochemical material of unexceptional structure. It lies entirely in the unique sensitivity of the sensory organs and brains of the insects.

One millimeter above the ground, where ants exist, things are radically different from what they seem to the gigantic creatures who peer down from a thousand times that distance. The ants do not follow the trail substance as a liquid trace on the soil, as we are prone to think. It comes up to them as a cloud of molecules diffusing through still air at the ground surface. The foragers move inside a long ellipsoidal space in which the gaseous material is dense enough to be detected. They sweep their paired antennae back and forth in advance of the head to catch the odorant molecules. The antennae are the primary sensory centers of the ant. Their surfaces are furred with thousands of nearly invisible hairs and pegs, among which are scattered diminutive plates and bottle-necked pits. Each of these sense organs is serviced by cells that carry electrical impulses into the central nerve of the antenna. Then relay cells take over and transmit the messages to the integrating regions of the brain. Some of the antennal organs react to touch, while others are sensitive to slight movements of air, so that the ant responds instantaneously whenever the nest is breached by intruders. But most of the sensors monitor the chemicals that swirl around the ant in combinations that change through each second of its life. Human beings live in a world of sight and sound, but social insects exist primarily by smell and taste. In a word, we are audiovisual where they are chemical.

The oddness of the insect sensory world is illustrated by the swift sequence of events that occurs along the odor trail. When a forager takes a wrong turn to the left and starts to run away from the track, its left antenna breaks out of the odor space first and is no longer stimulated by the guiding substance. In a few thousandths of a second, the ant perceives the change and pulls back to the right. Twisting right and left in response to the vanishing molecules, it follows a tightly undulating course between the nest and tree. During the navigation it must also dodge moment by moment through a tumult of other runners. If you watch a foraging worker from a few inches away with the unaided eye, it seems to touch each passerby with its antennae, a kind of tactile probe. Slow-motion photography reveals that it is actually sweeping the tips of the antennae over parts of the other ant's body to smell it. If the surface does not present exactly the right combination of chemicals — the colony's unique odor signature — the ant attacks at once. It may simultaneously spray an alarm chemical from special glands located in the head capsule, causing others in the vicinity to rush to the site with their mandibles gaping.

An ant colony is organized by no more than ten or twenty such signals, most of which are chemical secretions leaked or sprayed from glands. The workers move with swiftness and precision through a life that human beings have come to understand only with the aid of mathematical diagrams and molecular formulas. We can also simulate the behavior. Computer technology has made it theoretically possible to create a mechanical ant that duplicates the observed activity. But the machine, if for some reason we chose to build one, would be the size of a small automobile, and even then I doubt if it would tell us anything new about the ant's inner nature.

At the end of the trail the burdened foragers rush down the nest hole, into throngs of nestmates and along tortuous channels that end near the water table fifteen feet or more below. The ants drop the leaf sections onto the floor of a chamber, to be picked up by workers of a slightly smaller size who clip them into fragments about a millimeter across. Within minutes still smaller ants take over, crush and mold the fragments into moist pellets, and carefully insert them into a mass of similar material. This mass ranges in size between a clenched fist and a human head, is riddled with channels, and resembles a gray cleaning sponge. It is the garden of the ants: on its surface a symbiotic fungus grows which, along with the leaf sap, forms the ants' sole nourishment. The fungus spreads like a white frost, sinking its hyphae into the leaf paste to digest the abundant cellulose and proteins held there in partial solution.

The gardening cycle proceeds. Worker ants even smaller than those just described pluck loose strands of the fungus from places of dense growth and plant them onto the newly constructed surfaces. Finally, the very smallest—and most abundant—workers patrol the beds of fungal strands, delicately probing them with their antennae, licking their surfaces clean, and plucking out the spores and hyphae of alien species of mold. These colony dwarfs are able to travel through the narrowest channels deep within the garden masses. From time to time they pull tufts of fungus loose and carry them out to feed their larger nestmates.

The leafcutter economy is organized around this division of labor based on size. The foraging workers, about as big as houseflies, can slice leaves but are too bulky to cultivate the almost microscopic fungal strands. The tiny gardener workers, somewhat smaller than this printed letter I, can grow the fungus but are too weak to cut the leaves. So the ants form an assembly line, each successive step being performed by correspondingly smaller workers, from the collection of pieces of leaves out of doors to the manufacture of leaf paste to the cultivation of dietary fungi deep within the nest.

The defense of the colony is also organized according to size. Among the scurrying workers can be seen a few soldier ants, three hundred times heavier than the gardener workers. Their sharp mandibles are powered by massive adductor muscles that fill the swollen, quarter-inch-wide head capsules. Working like miniature wire clippers, they chop enemy insects into pieces and easily slice through human skin. These behemoths are especially adept at repelling large invaders. When entomologists digging into a nest grow careless, their hands become nicked all over as if pulled through a thorn bush. I have occasionally had to pause to stanch the flow of blood from a single bite, impressed by the fact that a creature one-millionth my size could stop me with nothing but its jaws.

No other animals have evolved the ability to turn fresh vegetation into mushrooms. The evolutionary event occurred only once, millions of years ago, somewhere in South America. It gave the ants an enormous advantage: they could now send out specialized workers to collect the vegetation while keeping the bulk of their populations safe in subterranean retreats. As a result, all of the different kinds of leafcutters together, comprising fourteen species in the genus *Atta* and twenty-three in *Acromyrmex,* dominate a large part of the American tropics. They consume more vegetation than any other group of animals, including the more abundant forms of caterpillars, grasshoppers, birds, and mammals. A single colony can strip an orange tree or bean patch overnight, and the combined populations inflict over a billion dollars' worth of damage yearly. It was with good reason that the early Portuguese settlers called Brazil the Kingdom of the Ants.

At full size, a colony contains three to four million workers and occupies three thousand or more underground chambers. The earth it excavates forms a pile twenty feet across and three to four feet high. Deep inside the nest sits the mother queen, a giant insect the size of a newborn mouse. She can live at least ten years and perhaps as long as twenty. No one has had the persistence to determine the true longevity. In my laboratory I have an individual collected in Guyana fourteen years ago. When she reaches eighteen, and breaks the proved longevity record of the seventeen-year locusts, my students and I will open a bottle of champagne to celebrate. In her lifetime an individual can produce over twenty million offspring, which translates into the following: a mere three hundred ants, a small fraction of the number emerging from a single colony in a year, can give birth to more ants than there are human beings on Earth.

The queen is born as a tiny egg, among thousands laid daily by the old mother queen. The egg hatches as a grublike larva, which is fed and laved incessantly throughout its month-long existence by the adult worker nurses. Through some unknown treatment, perhaps a special diet controlled by the workers, the larva grows to a relatively huge size. She then transforms into a pupa, whose waxy casement is shaped like an adult queen in fetal position, with legs, wings, and antennae folded tightly against the body. After several weeks the full complement of adult organs develops within this cuticle, and the new queen emerges. From the beginning she is fully adult and grows no more in size. She also possesses the same genes as her sisters, the colony workers. Their smaller size and pedestrian behavior is not due to heredity but rather to the different treatment they received as larvae.

In bright sunshine following a heavy rain, the virgin queen comes to the surface of the nest and flies up into the air to join other queens and the darkly pigmented, big-eyed males. Four or five males seize and inseminate her in quick succession, while she is still flying through the air. Their sole function now completed, they die within hours without returning to the home nest. The queen stores their sperm in her spermatheca, a tough muscular bag located just above and behind her ovaries. These reproductive cells live like independent microorganisms for years, passively waiting until they are released into the oviduct to meet an egg and create a new female ant. If the egg passes through the oviduct and to the outside without receiving a sperm, it produces a male. The queen can control the sex of her offspring, as well as the number of new workers and queens she produces, by opening or shutting the passage leading from her sperm-storage organ to the oviduct.

The newly inseminated queen descends to the ground. Raking her legs forward, she breaks off her wings, painlessly because they are composed of dead, membranous tissue. She wanders in a random pattern until she finds a patch of soft, bare soil, then commences to excavate a narrow tunnel straight down. Several hours later, when the shaft has been sunk to a depth of about ten inches, the queen widens its bottom into a small room. She is now set to start a garden and a colony of her own. But there is a problem in this life-cycle strategy. The queen has completely separated herself from the mother colony. Where can she obtain a culture of the vital symbiotic fungus to start the garden? Answer: she has been carrying it all along in her mouth. Just before leaving home, the young queen gathered a wad of fungal strands and inserted it into a pocket in the floor of her oral cavity, just back of the tongue. Now she passes the pellet out onto the floor of the nest and fertilizes it with droplets of feces.

As the fungus proliferates in the form of a whitish mat, the queen lays eggs on and around its surface. When the young larvae hatch, they are fed with other eggs given to them by the queen. At the end of their development, six weeks later, they transform into small workers. These new adults quickly take over the ordinary tasks of the colony. When still only a few days old, they proceed to enlarge the nest, work the garden, and feed the queen and larvae with tufts of the increasingly abundant fungus. In a year the little band has expanded into a force of a thousand workers, and the queen has ceased almost all activity to become a passive eating and egg-laying machine. She retains that exclusive role for the rest of her life. The measure of her Darwinian success is whether some of her daughters born five or ten years down the line grow into queens, leave on nuptial flights, and—rarest of all achievements—found new colonies of their own. In the world of the social insects, by the canons of biological organization, colonies beget colonies; individuals do not directly beget individuals.

People often ask me whether I see any human qualities in an ant colony, any form of behavior that even remotely mimics human thought and feeling. Insects and human beings are separated by more than 600 million years of evolution, but a common ancestor did exist in the form of one of the earliest multicellular organisms. Does some remnant of psychological continuity exist across that immense phylogenetic gulf? The answer is that I open an ant colony as I would the back of a Swiss watch. I am enchanted by the intricacy of its parts and the clean, thrumming precision. But I never see the colony as anything more than an organic machine.

Let me qualify that metaphor. The leafcutter colony is a superorganism. The queen sits deep in the central chambers, the vibrant growing tip from which all the workers and new queens originate. But she is not in any sense the leader or the repository of an organizational blueprint. No command center directs the colony. The social master plan is partitioned into the brains of the all-female workers, whose separate programs fit together to form a balanced whole. Each ant automatically performs certain tasks and avoids others according to its size and age. The superorganism's brain is the entire society; the workers are the crude analogue of its nerve cells. Seen from above and at a distance, the leafcutter colony resembles a gigantic amoeba. Its foraging columns snake out like pseudopods to engulf and shred plants, while their stems pull the green pieces down holes into the fungus gardens. Through a unique step in evolution taken millions of years ago, the ants captured a fungus, incorporated it into the superorganism, and so gained the power to digest leaves. Or perhaps the relation is the other way around: perhaps the fungus captured the ants and employed them as a mobile extension to take leaves into the moist underground chambers.

In either case, the two now own each other and will never pull apart. The ant-fungus combination is one of evolution's master clockworks, tireless, repetitive, and precise, more complicated than any human invention and unimaginably old. To find a colony in the South American forest is like coming upon some device left in place ages ago by an extra-terrestrial visitor for a still undisclosed purpose. Biologists have only begun to puzzle out its many parts.

Because of modern science the frontier is no longer located along the retreating wall of the great rain forest. It is in the bodies and lives of the leafcutters and thousands of other species found for the most part on the other side of that tragic line.

The Time Machine

YOU CAN ENVISION the full sweep of biology best by imagining that you own a motion-picture projector of magical versatility. The image it projects can be slowed to explode seconds into hours and days or speeded up to condense years and centuries into a few minutes. The image can be magnified to reveal microscopic detail or compressed to take in broad vistas from a distance. The projector serves as the scientist's time machine. It performs what Einstein called thought experiments.

Begin with a moment in history, any moment. Appropriate to our theme is the late evening of May 12, 1859, when Louis Agassiz and Benjamin Peirce are strolling in the spring air along a street in Cambridge, Massachusetts, conversing on the war between France and Austria and the threat to Switzerland's neutrality. It is a notable pair. Agassiz is the most celebrated American scientist of his generation, pioneer in the study of glaciers, leading authority on fishes and the general classification of animals, much-sought-after lecturer, professor at Harvard, founder of the Museum of Comparative Zoology, close friend of Emerson, Longfellow, and other literati, and about to become the most bitter and effective American opponent of Charles Darwin's theory of evolution. Peirce is a prominent mathematician and professor of astronomy at Harvard, already shaping up into one of Agassiz's strongest allies in the country's young intellectual community. The two are returning from a dinner given at the home of Asa Gray, professor of botany and Darwin's principal American supporter. The occasion was one of the fortnightly meetings of the Cambridge Scientific Club, consisting of about a dozen members of the Harvard faculty and others in the town with more than a passing interest in science. The event was one of the few that can be correctly called historic in the world of ideas: Gray has just presented the essentials of the Darwinian theory for the first time in the western hemisphere. Earlier in the year, he and Agassiz had circled each other warily at a meeting of the American Academy of Arts and Sciences, also held in Cambridge, sparring lightly on the vicarious distribution of plant species and other evidences of evolution but without addressing the central issue of the evolutionary process itself. Gray was too cautious to argue Darwin's theory of natural selection in a forthright manner before a large group of scholars with the formidable and popular Agassiz sitting there. Now, in the more relaxed company of the Cambridge Scientific Club, he has done so.

Few at the meeting sensed the importance of the Darwinian theory. The issue was cut between the two men alone. Gray enjoyed himself, letting the ideas and evidence tumble out in good spirits. Agassiz was disturbed. He said: "Gray, we must stop this." Indeed, much of the remainder of his career was spent trying to do just that. Now, at the moment on which we have chosen to focus the time machine, the conversation turns to current events, to war in Europe—a polite diversion from harder subjects about which close friends cannot afford to disagree. The historian A. Hunter Dupree has said of the two strollers, "Did they know that they stood on the knife-edge between two epochs in the intellectual history of the Western world? Did they know that the hesitant, eager tones of plain, familiar Asa Gray carried a message of more importance than Napoleon III, Franz Joseph, and all their legions rolled together?"

Picture the motion of the two men and the quiet flow of their words. Your effort consumes several seconds of time. Agassiz and Peirce and we, and all larger organisms, live in such *organismic time*, where most critical actions cover seconds or minutes. The reason for this deceptively simple fact is that human beings are constructed of billions of cells that communicate across their membranes by means of chemical surges and electrical impulses. A sentence is spoken: "Agassiz, I am much concerned." In a millisecond—one thousandth of a second—the compressed air strikes Agassiz's eardrum and transfers its energy inward to a row of three bony levers, which relay it instantly to the inner ear, an organ shaped like a snail shell; rows of sensitive cells deployed across the spiral resonate to the varying pitch of the vibration and trigger the discharge of equivalent nerve cells leading into the auditory nerve; as more milliseconds tick away, the coded electrical signals race into the hindbrain, cascade into predetermined pathways of the midbrain, the auditory cortex of the forebrain, and finally the seats of consciousness of the cerebral cortex—and Agassiz hears the sentence spoken by Peirce. Coordinated pulses of the neurons change their pattern through the cerebral cortex and special memory and emotive centers of the limbic system, generating new and quickly changing linkages of concepts and words; Agassiz is thinking. The brain combines new information from the banks of long-term memory into the temporary circuits of short-term memory. In a process consuming additional tenths of a second, the relevant images are pieced together and valuated by the emotive circuits they activate. Without pause the integrating centers of speech along the parietal cortex—Broca's and Wernicke's areas—are fully engaged, commands are issued through the cells of the relay stations of the motor cortex out to the tongue, lips, and larynx, and Agassiz responds: "Peirce, we must await developments." Four seconds have elapsed.

Now slow the reel in the time machine a thousand times. Agassiz and Peirce appear to freeze in their tracks. Their movements actually continue but are too slow to be perceived with unaided vision. Next magnify Agassiz until his individual nerve fibers come into view, then his cells, and finally molecules and atoms. Once again activity is normally paced and easily followed. The cell constituents swarm in passage through their appointed rounds, like the inhabitants of a city—like strollers in Cambridge. Enzyme molecules lock on to proteins and cleave them neatly into parts. A nerve cell discharges: along the length of its membrane, the voltage drops as sodium ions flow inward. At each point on the shaft of the nerve cell, these events consume several thousandths of a second, while the electrical signal they create—the voltage drop—speeds along the shaft at thirty feet per second. If we were to magnify the cell without slowing its action, the events would occur too swiftly to be seen. An electric discharge on the cell membrane would cross the field of vision faster than a rifle bullet. In order to understand such events at the molecular level, we must think in thousandths of seconds or less, the units of chemical reactions. That is why we slowed the action. We are in *biochemical time*. The magic projector allows us to picture its passage clearly by translating milliseconds into seconds, long enough for the myriad of our brain cells to interact and to recreate the image of the microscopic events that underlie the formation of the image now unfolding on the screen.

Spin the reel faster and return to organismic time. The biochemical reactions occur too swiftly to be comprehended, even if the magnification remains exactly right. So we reduce the magnification while moving back from Agassiz's body. As a consequence the projected atoms and molecules multiply and coalesce into vast aggregates, first as cells, then as tissues and organs. Once again, at these higher levels of biological organization, the action is slow enough to consume seconds—and hence to be understood by the brain. The diaphragm rises and falls, the heart pulsates, the leg muscles contract. Agassiz resumes his walk and conversation with Peirce.

Keep going. Speed the action still more—minutes and then hours pass within seconds—and pull away from Agassiz and Peirce. Like comic figures in an early silent movie, they speed jerkily out of the picture. As the reel turns ever faster, we rise above Cambridge to view the countryside of Massachusetts, then the full northeastern seaboard. Day and night pass in quickening succession. When the alternation between them reaches the flicker-fusion frequency, ten or more in a second of viewing time, they merge in our brains, so that the landscape is suffused by a continuous but dimmer light. Individual people and other organisms are no longer distinguishable except for a few long-lived trees that spring into existence and enlarge briefly before evaporating. But something new has appeared. We are aware of the presence of whole populations of species, say all of the sugar maples and red-eyed vireos, as they pass through cycles of expansion and retreat across the New England landscape. Ecosystems, formed of combinations of these species, have become the creatures of our vision. A pond is fringed with larch, fills up with waterweed, and then congeals into a bog. A sand dune sprouts beach grass, then wild rose and other low shrubs, which yield to jack pine and finally hardwood forest. We have entered *ecological time*. Biochemical events have been compressed beyond reckoning. Organisms are no more than ensembles defined by the mathematical laws of birth and death, competition, and replacement.

Where have Agassiz and Peirce and the other organisms of 1859 gone, during the acceleration of time? Dissolved into the gene pool of their species. Broken into tiny fragments by the shearing action of meiosis and fertilization. Erased as individuals, but preserved in perpetuity as DNA. They contributed half the genes of each of their children, one fourth of each grandchild, one eighth of each great-grandchild. Balancing this attrition was the multiplication of descendants through each successive generation. In a steady-state population, the average person has twice as many grandchildren as children, four times as many great-grandchildren, and so on up in a geometric progression. So the genes of one individual diffuse steadily outward through the population. Across a thousand years, the approximate threshold interval of *evolutionary time*, individuals lose most of their relevance as biological units. Families divide into multiplying lines of descent until they become coextensive across a large part of the population. Racial distinctions are blurred and eventually rendered meaningless. In the course of a thousand years, populations are even capable of splitting into entirely new species—although they have not done so in the case of the human line since the dawn of *Homo sapiens* half a million years ago.

The modern biological vision sweeps from microseconds to millions of years and from micrometers to the biosphere. But it is merely ordinary vision expanded by the electron microscope, earth-scan satellite, and other prosthetic devices of science and technology. The precise discipline is defined by the point of entry. Organismic biology explores the way we walk and speak; cell biology, the assembly and structure of our tissues; molecular biology, the ultimate chemical machinery; and evolutionary biology, the genetic history of our whole species. The modes of study depend upon the levels of organization chosen, which ascend in a hierarchical fashion: molecules compose cells, cells tissues, tissues organisms, organisms populations, and populations ecosystems. To understand any given species and its evolution requires a knowledge of each of the levels of organization sufficient to account for the one directly above it. Molecular biology is at the bottom (as its practitioners are always keen to point out) because everything depends upon the ultramicroscopic building blocks. Yet molecular biology on its own is a helpless giant. It cannot specify the parameters of space, time, and history that are crucial to and define the higher levels of organization. Consider the elementary fact that an embryo's development depends not just on its genes but on the way its cells are deployed in the surrounding environment. Or that an organism's behavior is shaped in part by learning, in other words by the alteration of its nerve cells by external stimuli. In a still deeper dimension, the very genes that comprise the central interest of molecular biology were assembled through a long history of mutation and selection within changing environments. When this last relationship became too obvious to ignore any longer, in the 1970s, molecular and evolutionary biology began to fuse, and the other branches of biology were realigned accordingly. The Darwinian conception approached its high watermark, more than a century after the publication of the *Origin of Species*.

AGASSIZ'S APPREHENSION over Darwin's theory deepened as he began to read *On the Origin of Species* late in 1859. "Agassiz, when I saw him last, had read but part of it," Asa Gray wrote J. D. Hooker in England in January 1860. "He says it is *poor—very poor*!! *(entre nous)*. The fact is that he is very much annoyed by it . . . Tell Darwin all this."

The impact of the *Origin* began to exceed that of Agassiz's own masterwork, the "Essay on Classification" in volumes of *Contributions to the Natural History of the United States*. The American zoologist had promoted his own theory of the origin of species: they are creations in the mind of God, brought to life when the creator thinks of them and extinguished when he ceases to think of them. It seemed a perfect conception, uniting science and religion in a form consistent with the transcendentalist beliefs then ruling America's intellectual scene. Its rejection by the Darwinians was something that Agassiz could not understand, no matter how hard he tried. Near the end of his life he complained:

> What of it, if it were true? Have those who object to repeated acts of creation ever considered that no progress can be made in knowledge without repeated acts of thinking? And what are thoughts but specific acts of mind? Why should it then be unscientific to infer that the facts of nature are the result of a similar process, since there is no evidence of any other cause?

Darwin wrote Asa Gray: "Agassiz's name, no doubt, is a heavy weight against us." But not his logic and evidence, which in letters to friends Darwin dismissed as wild, paradoxical, and religiously inspired. When Agassiz composed an article on the geology of the Amazon with arguments against evolution, Darwin told Charles Lyell he was glad to read it but "chiefly as a psychological curiosity."

Agassiz and Darwin were type specimens in a fundamental classification that far exceeded their conflict and history. There have always been two kinds of scientists, two kinds of natural philosophers. The first look upon the Creator, or at least the ineffability of the human spirit, as the ultimate explanation of first choice. The second follow the venerable dictum attributed to Polybius that, whenever it is possible to find out the cause of what is happening, one should not have recourse to the gods. The historian Loren Graham has given names to the two camps: restrictionists and expansionists. The first believe that science can go only so far, after which new forms of explanation and understanding have to be devised. The second acknowledge no intrinsic limits. They favor Bertrand Russell's definition of science as the things we know, distinct from philosophy as the things we do not know.

Darwin was the great expansionist. He shocked the world by arguing convincingly that life is the creation of an autonomous process so simple that it can be understood with just a moment of reflection. No equations, photons, or computer read-outs required. It can all be summarized in a couple of lines: new variations in the hereditary material arise continuously, some survive and reproduce better than others, and as a result organic evolution occurs. And even more briefly as follows: natural selection acting on mutations produces evolution. Given enough time (and the Earth is over four billion years old) even radically new kinds of organisms can be assembled this way, insects from myriapods, amphibians from lungfish, birds from small dinosaurs, and even life itself from inanimate matter.

Such a proposition was shocking in 1859 because before then almost everyone had worked under the opposite assumption, that great effects imply great causes. The eye of the eagle, the human hand, the whale's giant heart—such feats of engineering are so extraordinary as to suggest a designing power, if not God then at least an Idea of divine profundity. It had been difficult to think of the world in any other way. But Darwin showed that even the most complicated organism can be self-assembled through a series of small steps. God—and, yes, philosophy—was excused from the living world so that biology might seek its independent destiny.

And if that much were to be granted, what about the mind itself? The brain can also evolve by natural selection. If the mind is the creation of the brain, then it must be subject to material explanation. In 1838, shortly after he had conceived the principle of natural selection, Darwin wrote in his *N* Notebook that "to study Metaphysics, as they have always been studied, appears to me to be like puzzling at astronomy without mechanics.—Experience shows that the problem of the mind cannot be solved by attacking the citadel itself."

It certainly seemed to follow. The mind cannot understand its own workings and ultimate meaning merely by thinking about itself. If truly material in origin, the citadel is not to be entered directly but roundabout, through an exploration of the brain. The brain must lose some of its magic when it is regarded as a product of evolution through natural selection, like other organs of the body. So Darwin wrote in his 1838 *M* Notebook: "Origin of man now proved.—Metaphysics must flourish.—He who understand[s] baboon would do more toward metaphysics than Locke."

MODERN BIOLOGY has been built upon two great ideas. The first, a product of the nineteenth century, is that all life descended from elementary, single-celled organisms by means of natural selection. The second, perfected in the twentieth century, is that organisms are entirely obedient to the laws of physics and chemistry. No extraneous "vital force" runs the living cell. Each of the two ideas supports the other in compelling fashion. On the one side, the argument that organisms are physicochemical entities makes the universal operation of natural selection all the more plausible. On the other hand, the proof of natural selection in even a limited number of cases helps to explain why organisms are physicochemical mechanisms rather than the vessels of a mystic life force.

That is why expansionism has prevailed so far, passing beyond the boundaries of physics and chemistry into the domain of life and the mind, enabling its proponents to crank out knowledge at an accelerating rate. The biologist's time machine has grown into an awesome device that searches across centuries and down inside molecules. The vistas it has opened are the enchanted terrain of the new age.

But wait— a *machine*? Opening virgin territories? That has a familiar ring, and indeed we have arrived at the core of the fear of science, the cause of its historic alienation from the humanities. Science rampant is resisted as "scientism," an unpleasant doctrine. The dilemma of the machine in the garden exists in the realm of the spirit as surely as it does in the shrinking wilderness.

"Science grows and Beauty dwindles," wrote Tennyson. In the 1800s the romantic movement in poetry came to flourish as a fierce assault by free minds against the philosophy of the Enlightenment. They rejected the idea that all nature and human affairs are open to rational investigation, or that Newtonian law can be spread beyond physics. Keats warned in *Lamia:* "Philosophy will clip an Angel's wings. / Conquer all mysteries by rule and line, / Empty the haunted air, and gnomed mine— / Unweave a rainbow."

The romantic world view has been kept alive in sophisticated arguments by such modern theologians and philosophers as John Bowker, Theodore Roszak, and William Irwin Thompson. Their bill of indictment can be summarized:

"Science reduces and oversimplifies / Condenses and abstracts, drives toward generality / Presumes to break the insoluble / Forgets the spirit / Imprisons the spark of artistic genius."

The distinction between the two cultures of science and the humanities made famous by C. P. Snow thus persists. Until that fundamental divide is closed or at least reconciled in some congenial manner, the relation between man and the living world will remain problematic.

The Bird of Paradise

COME WITH ME NOW to another part of the living world. The role of science, like that of art, is to blend exact imagery with more distant meaning, the parts we already understand with those given as new into larger patterns that are coherent enough to be acceptable as truth. The biologist knows this relation by intuition during the course of field work, as he struggles to make order out of the infinitely varying patterns of nature.

Picture the Huon Peninsula of New Guinea, about the size and shape of Rhode Island, a weathered horn projecting from the northeastern coast of the main island. When I was twenty-five, with a fresh Ph.D. from Harvard and dreams of physical adventure in far-off places with unpronounceable names, I gathered all the courage I had and made a difficult and uncertain trek directly across the peninsular base. My aim was to collect a sample of ants and a few other kinds of small animals up from the lowlands to the highest part of the mountains. To the best of my knowledge I was the first biologist to take this particular route. I knew that almost everything I found would be worth recording, and all the specimens collected would be welcomed into museums.

Three days' walk from a mission station near the southern Lae coast brought me to the spine of the Sarawaget range, 12,000 feet above sea level. I was above treeline, in a grassland sprinkled with cycads, squat gymnospermous plants that resemble stunted palm trees and date from the Mesozoic Era, so that closely similar ancestral forms might have been browsed by dinosaurs 80 million years ago. On a chill morning when the clouds lifted and the sun shone brightly, my Papuan guides stopped hunting alpine wallabies with dogs and arrows, I stopped putting beetles and frogs in bottles of alcohol, and together we scanned the rare panoramic view. To the north we could make out the Bismarck Sea, to the south the Markham Valley and the more distant Herzog Mountains. The primary forest covering most of this mountainous country was broken into bands of different vegetation according to elevation. The zone just below us was the cloud forest, a labyrinth of interlocking trunks and branches blanketed by a thick layer of moss, orchids, and other epiphytes that ran unbroken off the tree trunks and across the ground. To follow game trails across this high country was like crawling through a dimly illuminated cave lined with a spongy green carpet.

A thousand feet below, the vegetation opened up a bit and assumed the appearance of typical lowland rain forest, except that the trees were denser and smaller and only a few flared out into a circle of blade-thin buttresses at the base. This is the zone botanists call the mid-mountain forest. It is an enchanted world of thousands of species of birds, frogs, insects, flowering plants, and other organisms, many found nowhere else. Together they form one of the richest and most nearly pure segments of the Papuan flora and fauna. To visit the mid-mountain forest is to see life as it existed before the coming of man thousands of years ago.

The jewel of the setting is the male Emperor of Germany bird of paradise (*Paradisaea guilielmi*), arguably the most beautiful bird in the world, certainly one of the twenty or so most striking in appearance. By moving quietly along secondary trails you might glimpse one on a lichen-encrusted branch near the tree tops. Its head is shaped like that of a crow—no surprise because the birds of paradise and crows have a close common lineage—but there the outward resemblance to any ordinary bird ends. The crown and upper breast of the bird are metallic oil-green and shine in the sunlight.

The back is glossy yellow, the wings and tail deep reddish maroon. Tufts of ivory-white plumes sprout from the flanks and sides of the breast, turning lacy in texture toward the tips. The plume rectrices continue on as wirelike appendages past the breast and tail for a distance equal to the full length of the bird. The bill is blue-gray, the eyes clear amber, the claws brown and black.

In the mating season the male joins others in leks, common courtship arenas in the upper tree branches, where they display their dazzling ornaments to the more somberly caparisoned females. The male spreads his wings and vibrates them while lifting the gossamer flank plumes. He calls loudly with bubbling and flutelike notes and turns upside down on the perch, spreading the wings and tail and pointing his rectrices skyward. The dance then reaches a climax as he fluffs up the green breast feathers and opens out the flank plumes until they form a brilliant white circle around his body, with only the head, tail, and wings projecting beyond. The male sways gently from side to side, causing the plumes to wave gracefully as if caught in an errant breeze. Seen from a distance his body now resembles a spinning and slightly out-of-focus white disk.

This improbable spectacle in the Huon forest has been fashioned by millions of generations of natural selection in which males competed and females made choices, and the accouterments of display were driven to a visual extreme. But this is only one trait, seen in physiological time and thought about at a single level of causation. Beneath its plumed surface, the Emperor of Germany bird of paradise possesses an architecture culminating an ancient history, with details exceeding those that can be imagined from the naturalist's simple daylight record of color and dance.

Consider one such bird for a moment in the analytic manner, as an object of biological research. Encoded within its chromosomes is the developmental program that led with finality to a male *Paradisaea guilielmi*. The completed nervous system is a structure of fiber tracts more complicated than any existing computer, and as challenging as all the rain forests of New Guinea surveyed on foot. A microscopic study will someday permit us to trace the events that culminate in the electric commands carried by the efferent neurons to the skeletal-muscular system and reproduce, in part, the dance of the courting male. This machinery can be dissected and understood by proceeding to the level of the cell, to enzymatic catalysis, microfilament configuration, and active sodium transport during electric discharge. Because biology sweeps the full range of space and time, there will be more discoveries renewing the sense of wonder at each step of research. By altering the scale of perception to the micrometer and millisecond, the laboratory scientist parallels the trek of the naturalist across the land. He looks out from his own version of the mountain crest. His spirit of adventure, as well as personal history of hardship, misdirection, and triumph, are fundamentally the same.

Described this way, the bird of paradise may seem to have been turned into a metaphor of what humanists dislike most about science: that it reduces nature and is insensitive to art, that scientists are conquistadors who melt down the Inca gold. But bear with me a minute. Science is not just analytic; it is also synthetic. It uses artlike intuition and imagery. In the early stages, individual behavior can be analyzed to the level of genes and neurosensory cells, whereupon the phenomena have indeed been mechanically reduced. In the synthetic phase, though, even the most elementary activity of these biological units creates rich and subtle patterns at the levels of organism and society. The outer qualities of *Paradisaea guilielmi*, its plumes, dance, and daily life, are functional traits open to a deeper understanding through the exact description of their constituent parts. They can be redefined as holistic properties that alter our perception and emotion in surprising and pleasant ways.

There will come a time when the bird of paradise is reconstituted by the synthesis of all the hard-won analytic information. The mind, bearing a newfound power, will journey back to the familiar world of seconds and centimeters. Once again the glittering plumage takes form and is viewed at a distance through a network of leaves and mist. Then we see the bright eye open, the head swivel, the wings extend. But the familiar motions are viewed across a far greater range of cause and effect. The species is understood more completely; misleading illusions have given way to light and wisdom of a greater degree. One turn of the cycle of intellect is then complete. The excitement of the scientist's search for the true material nature of the species recedes, to be replaced in part by the more enduring responses of the hunter and poet.

What are these ancient responses? The full answer can only be given through a combined idiom of science and the humanities, whereby the investigation turns back into itself. The human being, like the bird of paradise, awaits our examination in the analytic-synthetic manner. As always by honored tradition, feeling and myth can be viewed at a distance through physiological time, idiosyncratically, in the manner of traditional art. But they can also be penetrated more deeply than ever was possible in the prescientific age, to their physical basis in the processes of mental development, the brain structure, and indeed the genes themselves. It may even be possible to trace them back through time past cultural history to the evolutionary origins of human nature. With each new phase of synthesis to emerge from biological inquiry, the humanities will expand their reach and capability. In symmetric fashion, with each redirection of the humanities, science will add dimensions to human biology.

The Poetic Species

ONE OF THE most dramatic events of this century was the setting of the first Viking probe on Mars. The landing was scheduled for July 4, 1976, to coincide with the bicentennial of the United States, but it was actually accomplished on July 20. Scientists have anticipated few events with such electric excitement. There was an outside chance that Martian organisms might be detected quickly; a new biology could be created in one stroke. I know that feeling was shared by many following the event in the news, but I had a somewhat more personal interest. In 1964 I had attended a conference about Mars chaired by the ever ebullient Carl Sagan. We examined the best telescopic data available, speculating in every direction on the possible existence of life on the red planet and ways that it might be analyzed. I served as the instant "exo-ecologist," trying without much seriousness or effect to put a biological construction on the dark zones that spread and retreat through the middle latitudes (they were later proved to be sand storms). The conferees went home with no firm conclusions but high hopes for the NASA program then being charted.

Now at last, after twelve years of waiting, everyone was to have a close look at the surface of Mars, as though standing there in person, in a place where life might be found. The cameras scanned the Chryse Plain from the foot of the lander to the horizon and transmitted color pictures with a maximum resolution of less than a millimeter. The result was disappointing: no bushes dotted the landscape, no animals walked past the lens. A mechanical arm scooped up soil and analyzed it chemically to reveal reactions that were lifelike but not proof of the presence of microorganisms. The overall scene was nonetheless compelling: it was a landfall on another world remarkably like Earth in many respects. An outwardly familiar desert reached to the horizon where at sunset the thin atmosphere briefly glowed pink and turquoise. Every half-buried pebble, every wind pocket in the soil seen from a few feet away held the attention, hammered the imagination.

Then it was over. A giddying potential had been reduced to the merely known. The cold dust of the desert plain was committed to photographs in magazines, then to technical monographs, textbooks, and encyclopedias. The adventure became a set of facts, somehow mundane, to be looked up by students and recalled during leisure reading. A fundamental trait of science was exemplified: the magic had been consumed quickly and the action moved elsewhere. While a great deal remained to be learned, the high tide of research had swept on past an entire planet in less than a year.

Such brilliance fades quickly because newly discovered truths, and not truth in some abstract sense alone, are the ultimate goal and yardstick of the scientific culture. Scientists do not discover in order to know, they know in order to discover. That inversion of purpose is more than just a trait, it is the essence of the matter. Humanists are the shamans of the intellectual tribe, wise men who interpret knowledge and transmit the folklore, rituals, and sacred texts. Scientists are the scouts and hunters. No one rewards a scientist for what he knows. Nobel Prizes and other trophies are bestowed for the new facts and theories he brings home to the tribe. One great discovery and the scientist himself is great forever, no matter how foolish the rest of his deeds and pronouncements. No discovery, and he will probably be forgotten, even if he is learned and wise in matters scientific. The humanist grows in stature as he grows in wisdom. He can gain immortality as a critic, and justly so. But this vocational opportunity is not open—not yet—to the scientist. The most memorable critics among scientists are those who served as foils for the discoverers, helping them to clear error from the path. Thus the great Agassiz, cherished by Emerson and Longfellow, the idol of lecture halls along the Atlantic seaboard, is remembered most often today for having been wrong about Darwin.

Scientists therefore spend their productive lives struggling to reach the edge of knowledge in order to make discoveries. David Hilbert, the most successful mathematician of the early 1900s, stated the rules very well: "So long as a branch of science offers an abundance of problems, it will stay alive; a lack of problems foreshadows extinction or the cessation of independent development."

The scientist is not a very romantic figure. Each day he goes into the laboratory or field energized by the hope of a great score. He is brother to the prospector and treasure hunter. Every little discovery is like a gold coin on the ocean floor. The professional's real business, the bone and muscle of the scientific endeavor, amounts to a sort of puttering: trying to find a good problem, thinking up experiments, mulling over data, arguing in the corridor with colleagues, and making guesses with the aid of coffee and chewed pencils until finally something—usually small—is uncovered. Then comes a flurry of letters and telephone calls, followed by the writing of a short paper in an acceptable jargon. The great majority of scientists are hard-working, pleasant journeymen, not excessively bright, making their way through a congenial occupation.

Einstein spoke on the occasion of Max Planck's sixtieth birthday. He said that three types of people occupy the temple of science. There are those who enter for purely utilitarian reasons, to have a calling and invent things useful to mankind. Others are attracted by the sport in science, satisfying their ambition through the exercise of superior intellectual power. If an angel of the Lord were to come, Einstein said, and drive all belonging to these first two categories from the temple, a few people would be left, including Planck—"and that is why we love him."

The scientists most esteemed by their colleagues are those who are both very original and committed to the abstract ideal of truth in the midst of clamoring demands of ego and ideology. They pass the acid test of promoting new knowledge even at the expense of losing credit for it. They can face a fact, in accordance with the prayer of Thomas Henry Huxley, even though it slays them. Their principal aim is to discover natural law marked by *elegance,* the right mix of simplicity and latent power. The theory they accept is the one that defeats rival schemes by uniquely explaining the experiments of numerous independent investigators. It is a sleek instrument forged by repeated exposure to stubborn and sometimes inconvenient data. Conversely, the ideal experiment is the one that settles the rival claims of competing theories. Both the dominant theory and its patron data endure only if they fit the explanations of other disciplines through a network of logically tight and quantitative arguments.

This tidy conception is made the more interesting by the deep epistemological problem it creates and the biological process it implies. Elegance is more a product of the human mind than of external reality. It is best understood as a product of organic evolution. The brain depends upon elegance to compensate for its own small size and short lifetime. As the cerebral cortex grew from apish dimensions through hundreds of thousands of years of evolution, it was forced to rely on tricks to enlarge memory and speed computation. The mind therefore specializes on analogy and metaphor, on a sweeping together of chaotic sensory experience into workable categories labeled by words and stacked into hierarchies for quick recovery. To a considerable degree science consists in orginating the maximum amount of information with the minimum expenditure of energy. Beauty is the cleanness of line in such formulations, along with symmetry, surprise, and congruence with other prevailing beliefs. This widely accepted definition is why P. A. M. Dirac, after working out the behavior of electrons, could say that physical theories with some physical beauty are also the ones most likely to be correct, and why Hermann Weyl, the perfecter of quantum and relativity theory, made an even franker confession: "My work always tried to unite the true with the beautiful; but when I had to choose one or the other, I usually chose the beautiful."

Einstein offered the following solution to the dilemma of truth versus beauty: "God does not care about our mathematical difficulties. He integrates empirically." In other words, a mind with infinite memory store and calculating ability could compute any system as the sum of all its parts, however minute and numerous. Mathematics and beauty are devices by which human beings get through life with the limited intellectual capacity inherited by the species. Like a discerning palate and sexual appetite, these esthetic contrivances give pleasure. Put in more mechanistic terms, they play upon the circuitry of the brain's limbic system in a way that ultimately promotes survival and reproduction. They lead the scientist adventitiously into the unexplored fractions of space and time, from which he returns to report his findings and fulfill his social role. Riemannian geometry is declared beautiful no less than the bird of paradise, because the mind is innately prepared to receive its symmetry and power. Pleasure is shared, triumph ceremonies held, and the communal hunt resumed. In a memorial tribute to Hermann Minkowski, David Hilbert described this perpetual cycle with gentle botanic images:

> Our Science, which we loved above everything, had brought us together. It appeared to us as a flowering garden. In this garden there were wellworn paths where one might look around at leisure and enjoy oneself without effort, especially at the side of a congenial companion. But we also liked to seek out hidden trails and discovered many an unexpected view which was pleasing to

> our eyes; and when the one pointed it out to the other, and we admired it together, our joy was complete.

Scientific innovation sometimes sounds like poetry, and I would claim that it is, at least in the earliest stages. The ideal scientist can be said to think like a poet, work like a clerk, and write like a journalist. The ideal poet thinks, works, and writes like a poet. The two vocations draw from the same subconscious wellsprings and depend upon similar primal stories and images. But where scientists aim for a generalizing formula to which special cases are obedient, seeking unifying natural laws, artists invent special cases immediately. They transmit forms of knowledge in which the knower himself is revealed. Their works are lit by a personal flame and above all else they identify, in Roger Shattuck's expression, "the individual as the accountable agent of his action and as the potential seat of human greatness."

The aim of art is not to show how or why an effect is produced (that would be science) but literally to produce it. And not by just any cry from the heart—it requires mental discipline no less than in science. In poetry, T. S. Eliot explained, the often-quoted criterion of sublimity misses the mark. What counts is not the greatness of the emotion but the intensity of the artistic process, the pressure under which the fusion takes place. The great artist touches others in surgical manner with the generating impulse, transferring feeling precisely. His work is personal in style but general in effect.

Ideally art is powerful enough to cross cultures; it reads the code of human nature. Octavio Paz's poem "The Broken Waterjar" *(El cántaro roto)* accomplishes this result with splendid effectiveness. Paz is torn by the contradiction in the Mexican experience. He says that the minds of his people are capable of long flights of imagination and visions of piercing beauty. The people look at the sky and add torches, wings, and "bracelets of flaming islands." But they also look down to a desiccated landscape, symbolizing physical and spiritual poverty. A potentially great nation has been divided by the Conquest and stifled by oppression:

> Bare hills, a cold volcano, stone and a sound of panting
> under such splendor, and drouth, the taste of dust,
> the rustle of bare feet in the dust, and one tall tree in the
> middle of the field like a petrified fountain!

The resolution is not offered in the form of practical advice, which might easily prove wrong, but in the poet's vision of unity in a search back through time, "más allá de las aguas del bautismo," to a more secure metaphysical truth, and Paz says:

> vida y muerte no son mundos contrarios, somos un solo
> tallo con dos flores gemelas

Mexico is a single stem with twin flowers, united by the continuity of time.

The essence of art, no less than of science, is synecdoche. A carefully chosen part serves for the whole. Some feature of the subject directly perceived or implied by analogy transmits precisely the quality intended. The listener is moved by a single, surprising image. In "The Broken Waterjar" the rustle of bare feet in the dust conveys the pauperization of Mexico. The artist knows which sensibilities shared by his audience will permit the desired impact.

Picasso defined art as the lie that helps us to see the truth. The aphorism fits both art and science, since each in its own way seeks power through elegance. But this inspired distortion is only a technique of thinking and communication. There is a still more basic similarity: both are enterprises of discovery. And the binding force lies in our biology and in our relationship to other organisms. In art, the workings of the mind are explored, whereas in science the domain is the world at large and now, increasingly, the workings of the mind as well. Of equal importance, both rely on similar forms of metaphor and analogy, because they share the brain's strict and peculiar limitations in the processing of information.

Most scientists have been self-conscious at one time or another about their own procedures of discovery. The stakes are high: a major advance can be made by a single insight consuming just a few seconds. Scientific theory is the last and greatest of the cottage industries and the principal source of vitality in every discipline. Is there a secret, a hidden meta-formula by which the mind creates these visible formulas? This question has been addressed in studies of creativity by cognitive psychologists. Their work has advanced rapidly during the past ten years as the iron grip of behaviorist philosophy relaxed and studies of the mind became more respectable. Of equal weight is the testimony of scientists about their own steps to discovery. Essays by Freeman Dyson, J. B. S. Haldane, Werner Heisenberg, Willard Libby, Henri Poincaré, John Wheeler, Chen Ning Yang, and others form a veritable psychologist's case book in this most elusive of mental operations.

In my own search for a meta-formula I have had the good fortune of working with gifted mathematicians on subjects for which there was little or no previous theory—no framework of well-defined ideas on which information might be deployed and linked in explanatory chains. Early in my life I discovered that I have very little talent for mathematics. It is simply one of those things you either have or not, and no amount of training or effort will bring it to you, just as most people have little capacity for playing the violin or running a fast mile. Through hard work at college and while a young professor I became mathematically semiliterate. I can puzzle through articles on pure theory in the journals and advanced textbooks, but I cannot write the rows of original equations that transport the mind from one or two easy propositions by some miracle to a new, counterintuitive truth. My ability lies in seeing the problem in the first place, envisioning what a subject might look like if a proper theoretical scaffolding and beautiful facts were put in place. In other words scout, not architect. Nothing is more attractive to me than a muddled domain awaiting its first theory. I feel most at home with a jumble of glittering data and the feeling that they might be fitted together for the first time into some new pattern. This inclination made me especially compatible with mathematicians. I became fascinated with the way they think, why they should be so much better at quantitative reasoning than I, what difference it made in the end, why I should be the one so often to suggest moving in a particular direction, but then even more frequently not be able to do so, and finally how different everything looked after a little progress had been made.

From this personal experience and the impressions recorded by others, let me offer the following rough map of innovation in science. You start by loving a subject. Birds, probability theory, explosives, stars, differential equations, storm fronts, sign language, swallowtail butterflies—the odds are that the obsession will have begun in childhood. The subject will be your lodestar and give sanctuary in the shifting mental universe.

A pioneer in molecular biology (still young, because most pioneering work was done after 1950) once told me that his fascination with the replication of DNA molecules began when he was given an erector set as a child. Playing with the toy, he saw the possibilities of creation by the multiplication and rearrangement of identical units. The great metallurgist Cyril Smith owed his devotion to alloys to the fact that he was color blind. The impairment caused him to turn his attention at an early age to the intricate black-and-white patterns to be seen everywhere in nature, to swirls, filigree, and banding, and eventually to the fine structure of metal. Albert Camus spoke for all such innovators when he said that "a man's work is nothing but this slow trek to rediscover, through the detours of art, those two or three great and simple images in whose presence his heart first opened."

The subject we love is probably also well known to others. So we have to travel away from it into regions deliberately chosen for their lack of previous attention. Science has flourished in western cultures because this difficult step was recognized by society as valuable, and rewarded. Nothing comes harder than original thought. Even the most gifted scientist spends only a tiny fraction of his waking hours doing it, probably less than one tenth of one percent. The rest of the time his mind hugs the coast of the known, reworking old information, adding lesser data, giving reluctant attention to the ideas of others (what use can *I* make of them?), warming lazily to the memory of successful experiments, and looking for a problem—always looking for a problem, something that can be accomplished, that will lead somewhere, anywhere.

There is in addition an optimal degree of novelty in problem-seeking, difficult to measure and follow. Stick to the coast too tightly and only minor new data will follow. Venture out of sight and you risk getting lost at sea. Years of effort might then be wasted, competitors will hint that the enterprise is pseudoscience, grants and other patronage will be cut off, and tenure and election to the academies denied. The fate of the overly daring is to sail off the rim of the world.

On one point both psychologists and successful voyagers agree. The key instrument of the creative imagination is analogy. Hideki Yukawa, who reflected on this matter for forty years while working on the nuclear binding force, explained it as follows:

> Suppose that there is something which a person cannot understand. He happens to notice the similarity of this something to some other thing which he understands quite well. By comparing them he may come to understand the thing which he could not understand up to that moment. If his understanding turns out to be appropri-

ate and nobody else has ever come to such an understanding, he can claim that his thinking was really creative.

We have returned to the common, human origin of science and art. The innovator searches for comparisons that no one else has made. He scrambles to tighten his extension by argument, example, and experiment. Important science is not just any similarity glimpsed for the first time. It offers analogies that map the gateways to unexplored terrain. The comparisons meet the criterion of principal metaphor used by art critics: one commanding image synthesized from several units, such that a single complex idea is attained not by analysis but by the sudden perception of an objective relation.

Theoretical scientists, inching away from the safe and known, skirting the point of no return, confront nature with a free invention of the intellect. They strip the discovery down and wire it into place in the form of mathematical models or other abstractions that define the perceived relation exactly. The now-naked idea is scrutinized with as much coldness and outward lack of pity as the naturally warm human heart can muster. They try to put it to use, devising experiments or field observations to test its claims. By the rules of scientific procedure it is then either discarded or temporarily sustained. Either way, the central theory encompassing it grows. If the abstractions survive they generate new knowledge from which further exploratory trips of the mind can be planned. Through the repeated alternation between flights of the imagination and the accretion of hard data, a mutual agreement of the workings of the world is written, in the form of natural law.

The scientist entrepreneur can pick a subject virtually at random and soon be on the edge of discovery, if he is at all lucky. In 1962 Robert H. MacArthur and I, both in our early thirties, decided to try something new in biogeography. The discipline, which studies the distribution of plants and animals around the world, was ideal for theoretical research. Biogeography was intellectually important, replete with poorly organized information, underpopulated, and almost devoid of quantitative models. Its borders with ecology and genetics, specialties in which we also felt well prepared, were blank swaths across the map.

MacArthur was then an associate professor of biology at the University of Pennsylvania, the same rank I held at Harvard. He later moved to Princeton, where he spent the remainder of his short life. He was medium tall and thin, with a handsomely angular face. He met you with a level gaze supported by an ironic smile and widening of eyes. He spoke with a thin baritone voice in complete sentences and paragraphs, signaling his more important utterances by tilting his face slightly upward and swallowing. He had a calm understated manner, which in intellectuals suggests tightly reined power. Because very few professional academics can keep their mouths shut long enough to be sure about anything, MacArthur's restraint gave his conversation an edge of finality he did not intend. In fact he was basically shy and reticent. He was not a mathematician of the first class—very few scientists are, otherwise they would become pure mathematicians—but he joined superior talent in that field with an extraordinary creative drive, decent ambition, and a love of the natural world, birds, and science, in that order.

By general agreement MacArthur was the most important ecologist of his generation. His use of evolutionary theory in the explanation of population growth and competition was so original and productive that biologists today refer informally to the MacArthur school of ecology, or more justly to the Hutchinson-MacArthur school, in order to include his influential teacher at Yale, G. Evelyn Hutchinson. MacArthur died of renal cancer in 1972. Hours before he died in his sleep, I talked with him at length over the telephone, from Cambridge to Princeton. It was the same as ten years before. We touched on familiar subjects: the future of ecology, the key unsolved problems of evolution, and the merits of various colleagues. MacArthur's easy concentration on these matters, as if he had a hundred years to live, was but one more measure of his intellectual integrity.

In 1960, when we first met, I was finishing a ten-year stint of field work and knew the distribution of animals quite well. I had worked out the classification of hundreds of species of ants throughout the Pacific region and elsewhere. I had the sense that there was some general order within the exciting chaos, some powerful process to be uncovered, but only a vague idea of its outline. In our first laconic discussion (MacArthur had the effect on me of shortening my sentences), we quickly realized that something of value lay close to the surface. In the following exchange I have telescoped our conversations and letters on the subject in order to convey the crucial steps in the origin of species-equilibrium theory—almost, as it were, out of the air.

Wilson: I think biogeography can be made into a science. There are striking regularities no one has explained. For example, the larger the island, the more the species of birds or ants that live on it. Look at what happens when you go from little islands, such as Bali and Lombok, to big ones like Borneo and Sumatra. With every tenfold increase in area, there is roughly a doubling of the number of species found on the island. That appears to be true for most other kinds of animals and plants for which we have good data. Here's another piece in the puzzle. I've found that as new ant species spread out from Asia and Australia onto the islands between them, such as New Guinea and Fiji, they eliminate other ones that settled there earlier. At the level of the species this fits in pretty well with the views of Philip Darlington and George Simpson. They proved that in the past major groups of mammals, such as all the deer or all the pigs taken together, have tended to replace other major groups in South America and Asia, filling the same niches. So there seems to be a balance of nature down to the level of the species, with waves of replacement spreading around the world.

MacArthur: Yes, a species equilibrium. It looks as

though each island can hold just so many species, so if one species colonizes the island, an older resident has to go extinct. Let's treat the whole thing as if it were a kind of physical process. Think of the island filling up with species from an empty state up to the limit. That's just a metaphor, but it might get us somewhere. As more species settle, the rate at which they are going extinct will rise. Let me put it another way: the probability that any given species will go extinct increases as more species crowd onto the island. Now look at the species arriving. A few colonists of each are making it each year on the wind or floating logs or, like birds, flying in on their own power. The more species that settle on the island, the fewer *new* ones that will be arriving each year, simply because there are fewer that aren't already there. Here's how a physicist or economist would represent the situation. As the island fills up, the rate of extinction goes up and the rate of immigration goes down, until the two processes reach the same level. So by definition you have a dynamic equilibrium. When extinction equals immigration, the *number* of species stays the same, even though there may be a steady change in the particular species making up the fauna.

Look what happens when you play around a little with the rising and falling curves. Let the islands get smaller. The extinction rates have to go up, since the populations are smaller and more liable to extinction. If there are only ten birds of a kind sitting in the trees, they are more likely to go to zero in a given year than if there are a hundred. But the rate at which new species are arriving won't be affected very much, because islands well away from the mainland can vary a lot in size without changing much in the amount of horizon they present to organisms traveling toward them. As a result, smaller islands will reach equilibrium sooner and end up with a smaller number of species at equilibrium. Now look at pure distance as a factor. The farther the island is from the source areas, say the way Hawaii is farther out in the Pacific than New Guinea, the fewer new species that will be arriving each year. But the rate of extinction stays the same because, once a species of plant or animal is settled on an island, it doesn't matter whether the island is close or far. So you expect the number of species found on distant islands to be fewer. The whole thing is just a matter of geometry.

Weeks pass. We are sitting next to the fireplace in MacArthur's living room, with notes and graphs spread out on a coffee table.

Wilson: So far so good. The numbers of bird and ant species *do* go down as islands get smaller and farther from the mainland. We'll label the two trends the *area effect* and the *distance effect.* Let's take them both as given for the moment. How do we know that they prove the equilibrium model? I mean, other people are almost certainly going to come up with a rival theory to explain the area and distance effects. If we claim that the results prove the model that predicted them, we will commit what logicians call the Fallacy of Affirming the Consequent. The only way we can avoid that impasse is to get results that are uniquely predicted by our model and no one else's.

MacArthur: All right, we've gone this far with pure abstraction — let's go on. Try the following: line up the extinction and emigration curves so that where they cross and create the equilibrium, they are straight lines and tilted at approximately the same angle. As an exercise in elementary differential calculus, you can show that the number of years an island takes to fill up to 90 percent of its potential should just about equal the number of species at equilibrium divided by the number going extinct every year.

Wilson: Let's look at Krakatoa.

KRAKATOA is the small island in the Sunda Strait between Java and Sumatra that exploded with an equivalent force of approximately 100 megatons of TNT on August 27, 1883. As a wave raced out across the Indian Ocean (eventually to cause ships to tug at their anchor chains in the English Channel) a blanket of glowing hot pumice covered most of what was left of the island and killed the last remnants of life. Scientists realized that they had a once-in-a-century opportunity to witness the recolonization of a dead island. Between 1884 and 1936 several principal expeditions were mounted under the auspices of the Dutch colonial government to follow the return of plants and animals to Krakatoa. The data were published in a scattering of articles and books, but very little use had been made of them in the years that followed, largely because no quantitative theory of island biogeography existed.

The Dutch reported that the vegetation returned to Krakatoa quickly. The first plants sprouted in the rain-moistened ash within a year, and a luxuriant forest covered most of its surface by 1920. Large numbers of animal species simultaneously colonized the island. The journal data were especially good for birds. We got Krakatoa's area and fitted it on our own curve relating area to species. Krakatoa should have 30 bird species at equilibrium. The Dutch surveys indicated that it reached about 90 percent of that number in 30 years. The elementary equilibrium equation predicted that by the late 1920s about one species per year (30 species divided by 30 years) should be going extinct, to be replaced by one new species coming in. We scanned the pages of the reports eagerly. Would the Dutch scientists mention extinction? They did. They were impressed, they said, by a remarkably high turnover in bird species. We calculated that they saw an average of one extinction every five years. That rate was five times lower than our prediction, but nevertheless much higher than most naturalists expect to find on such islands. And when extinctions and immigrations occurring unseen in the intervals between the surveys were provisionally figured in, the fit to the theory was closer.

Other biologists were encouraged by this mathematically simple attempt to define a dynamic equilibrium. They were intrigued by the very idea that the diversity of life rises to a certain level and stays there, with species coming and going at a predictable rate. The theory could be applied not just to islands in the ocean but also to " habitat islands, " such as woodlots in a sea of grass, ponds and streams in a sea of land, and in fact to any habitat enclosed by a different environment hostile to its organisms. It might even be used to predict the fate of parks and nature reserves over a period of years or centuries.

Species-equilibrium theory, in other words, was heuristic. It promoted further study, a quality highly valued in science. In giving answers to a few old questions, it raised new ones and suggested techniques for their solution. New, more elaborate studies soon proliferated. Mountain tops, lakes, coral reefs, and bottles of water were added to the list of habitat islands under study, while guidelines were suggested for the design of parks. The World Wildlife Fund used some of the models in planning its rain-forest reserve project near Manaus. It was all very exciting, but the first models that MacArthur and I had fashioned were too crude to fit these additional cases. A whole new canon of theory was invented, and appropriate experiments followed. The study of species equilibria grew into a rich and sophisticated branch of ecology. Twenty years later our particular contributions were no longer clearly distinguishable from those of equal or superior merit contributed by other biologists. The surviving fragments had been absorbed into the mainstream, which continues to broaden and shift each year.

That is the way of science. The scientist may think like a poet, but the products of his imagination are seldom preserved in their original state. It is often said that a discipline is successful according to how quickly its founders are forgotten—or, more precisely, how soon they are replaced in the textbooks and vade mecums of the trade. No original MacArthurs hang in galleries; no biologists return to his original texts in the *Proceedings of the National Academy of Sciences* to sift for nuance and symbolism. MacArthur lives on as he would have liked, in the irreversible change he caused in an important branch of science.

AT THE MOMENT the spark ignites, when intuition and metaphor are all-important, the artist most closely resembles the scientist. But he does not then press on toward natural law and self-dissolution within the big picture. All his skills are aimed at the instant transference of images and control of emotions in others. For purposes of craft, he carefully avoids exact definitions or the display of inner logic.

In 1753 Bishop Lowth made the correct diagnosis in *Lectures on the Sacred Poetry of the Hebrews:* the poetic mind is not satisfied with a plain and exact description but seeks to heighten sensation. "For the passions are naturally inclined to amplifications; they wonderfully magnify and exaggerate whatever dwells upon the mind, and labour to express it in animated, bold, and magnificent terms."

The essential quality can be rephrased in more modern terms. The mind is biologically prone to discursive communication that expands thought. Mankind, in Richard Rorty's expression, is the poetic species. The symbols of art, music, and language freight power well beyond their outward and literal meanings. So each one also condenses large quantities of information. Just as mathematical equations allow us to move swiftly across large amounts of knowledge and spring into the unknown, the symbols of art gather human experience into novel forms in order to evoke a more intense perception in others. Human beings live—literally live, if life is equated with the mind—by symbols, particularly words, because the brain is constructed to process information almost exclusively in their terms.

I have spoken of art as a device for exploration and discovery. Its practitioners and expert observers, whose authority is beyond question, have stressed other functions as well. In Samuel Johnson's definition, to instruct by pleasing. According to Keats, to uplift by the refinement of shared feelings. No — moral, the role of art is moral, according to D. H. Lawrence. A spell against death, to create and preserve the self, in the formulation of Richard Eberhart. For the more prosaic cultural anthropologists, art above all else expresses the purposes of a society. Indeed, affirmation may have been the original evolutionary driving force behind Paleolithic cave art. It was certainly served by the early oral poets of Europe, including the illiterate Homeric bards who recited the *Iliad* and *Odyssey* at festivals and thus transmitted the central myths and legends of ancient Greece. When this cohesive function fails, and tradition and taste fragment as part of culture's advance, criticism becomes a necessary and honored profession. Then we also witness revolutionary art, which goes beyond innovation to promote a different society and culture.

All these functions are variously filled according to circumstances. Nevertheless, art generally considered to be important appears to be marked by one consistent quality: it explores the unknown reaches of the mind. The departure is both calculated and tentative, as in science.The poet focuses on the inward search itself and attracts us to his distant constructions. Something moves on the edge of the field of vision, a new connection is glimpsed, holds for a moment. Words pour in and around, and the image takes substantial form, at first believed familiar, then seen as strikingly new. It is something, as in Thomas Kinsella's "Midsummer,"

> that for this long year
> Had hid and halted like a deer
> Turned marvellous,
> Parted the tragic grasses, tame,
> Lifted its perfect head and came
> To welcome us.

But the poet refuses to take us any farther. If he goes on the precise image will melt into abstract descriptions; light and beauty will congeal into rows of formulas. In this essential way art differs from science. The world of interest is the mind, not the physical universe on which mental process feeds. Richard Eberhart, a keen observer of nature, listened to the same New England birds that led Robert MacArthur to mathematical theories of ecology, and I daresay that the first swirl of imagery, the first tensile pleasures were the same, but the two then diverged, the poet inward and the scientist outward into separate existences:

> No, may the thrush among our high pine trees
> Be ambiguous still, elusive in true song,
> Never or seldom seen, and if never seen
> May it to my imperious memory belong.

He holds back, on himself and on us, in order to cast his spell. Again we see that the dilemma of the machine-in-the-garden exists in the rain forests of the mind as surely as it does in the American continent. Our intrinsic emotions drive us to search for fresh habitats, to cross unexplored terrain, but we still crave the sense of a mysterious world stretching infinitely beyond. The free-living birds (thrush, nightingale, bird of paradise), being rulers of the blank spaces on the map and negligent of human existence, are worthy symbols of both art and science.

I have emphasized the expansive role of poetry to argue that, whereas art and science are basically different in execution, they are convergent in what they might eventually disclose about human nature. Until recently science was minimally concerned with the mind. Even those who granted mental process a material origin classified it as an ephemeron, the proper subject of some other occupation, a different way of thinking, a separate literature—in short, the humanities. Now all that has changed. Cognitive psychology has emerged as a strong discipline. Parallel studies on the nervous system and artificial intelligence are contributing further insights. Scientists view the human brain as one of the last remaining frontiers of empirical research. They are beginning to pour in from genetics, molecular biology, and other neighboring disciplines to join its settlement.

The subject of greatest immediate interest is long-term memory. We are essentially what we remember or can remember at some time in the future. People build memory by linking new images and concepts to old ones. In its size, in the space it fills, the mind expands like a coral reef, adding new branches and cross-ties out from the edge of those parts already established and anchored, while its central body settles and coalesces. A widely used theory of learning uses a more abstract but equivalent metaphor, the node-link model. The nodes are concepts such as dog, red, bark, aboveground, running, and teeth. Each node is linked to certain others, so that the activation of one in a person's memory tends to pull in a whole group. The image of a dog (or the mere word "dog")may evoke red, running, aboveground, fur, teeth, and a great deal more: concatenations of memories, frames of node-link structures riffled back and forth through time, and emotional nodes that can only be labeled by broad generic words such as fear and affection. The mind probes and adapts by a process that some psychologists have called spreading activation. Given new images or altered circumstances, it encompasses widening circles of nodes and links in a search for similarities, finally settling on the best categories and analogies previously stored in long-term memory.

Suppose that a strange new animal walks out of the tangled undergrowth and into our view. It might be compared to a dog, or a monkey, or something else. Perhaps the mind, overloaded with its novelty, will simply abandon the search, giving the creature a new name and a fuller description than usual in order to establish a new node-link cluster. *This dark furry beast—let us call it X—is smaller than a dog. It has batlike ears, round luminous eyes, and ratlike teeth. It creeps about sluggishly while picking at objects with long spidery fingers. At night it prowls through the treetops and inspires superstitious terror in the few natives who glimpse it by torchlight along the forest trails.* (This particular animal happens to be the aye-aye of Madagascar.)

SO NOW there can be a more explicit description of what theoretical scientists and artists, dreamers of a kind, accomplish during the first stages of original thought. It is controlled growth, a disciplined spread of the mind into hidden recesses where concepts and linkages are still embryonic or nonexistent.

Genius is this kind of expertise born aloft on the wings of energy, daring, and luck. The combination came together in a famous letter sent in 1913 from Srinivasa Ramanujan, a Hindu clerk, to Godfrey H. Hardy, the English mathematician. By the age of twenty-five and with the aid of only one obscure text on higher mathematics, Ramanujan had independently solved some of the problems that previously occupied the best mathematicians of Europe for over a century. A few of the equations were known already: number (1.8) in Ramanujan's series was a formula of Laplace first proved by Jacobi, while (1.9) had already been published by Rogers in 1907. Equations (1.5) and (1.6) looked somewhat familiar and could be confirmed, according to Hardy, but only with a surprising amount of effort. Then came something really new:

> The formulae (1.10)–(1.13) are on a different level and obviously both difficult and deep . . . (1.10)–(1.12) defeated me completely; I had never seen anything in the least like them before. A single look at them is enough to show that they could only be written down by a mathematician of the highest class. They must be true because, if they were not true, no one would have had the imagination to invent them.

They must be true: the same can be said of outstanding achievements in literature and the arts, which pull the rest of us along until the construction becomes self-evident. Eliot wrote that " unless we have those few men who combine an exceptional sensibility with an exceptional power over words, our own ability, not merely to express, but even to feel any but the crudest emotions, will degenerate." The difference in power is one of degree rather than kind, but it crosses a threshold to create a qualitative new result in the same way that a critical speed lifts a glider off the ground into flight.

Research on cognitive development has shown that in the course of its growth the mind probes certain channels much more readily than others. Some of the responses are automatic and can be measured by physiological changes of which the individual is mostly or entirely unaware. For example, using electroencephalograms in the study of response to graphic designs, the Belgian psychologist Gerda Smets found that maximal arousal (measured by the blockage of the alpha wave) occurs when the figure contains about 20 percent redundancy. That is the amount present in a spiral with two or three turns, or a relatively simple maze, or a neat cluster of ten or so triangles. Less arousal occurs when the figure consists of only one triangle or square, or when the design is more complicated than the optimum—as in a difficult maze or an irregular scattering of twenty rectangles. The data are not the result of a chance biochemical quirk. When selecting symbols and abstract art, people actually gravitate to about the levels of complexity observed in Smets's experiments. Furthermore, the preference has its roots in early life. Newborn infants gaze longest at visual designs containing between five and twenty angles. During the next three months their preference shifts toward the adult pattern measured with electroencephalograms. Nor is there anything foreordained or otherwise trivial in the aesthetic optimum of human beings. It is easy to imagine the evolution of some other intelligent species in another time or on some other planet, possessing different eyes, optic nerves, and brain—and thus distinct optimal complexity and artistic standards.

We can reasonably suppose that the compositions of artists play upon the rules of mental development that are now beginning to receive the objective attention of experimental psychology. The distinction between science and art can be understood more clearly from this different perspective. The abstracted qualities of the developmental rules of the mind are the principal concern of science. In contrast, the node-link structures themselves, their emotional color, tone, cadence, fidelity to personal experience, and the images they fleetingly reveal, are more the domain of art. Of equal importance to both enterprises are the symbols and myths that evoke the mental structures in compelling fashion. Certain great myths—the origin of the world, cataclysm and rebirth, the struggle between the powers of light and darkness, Earth Mother, and a few others—recur dependably in cultures around the world. Lesser, more personal myths appear in crisis poems and romantic tales, where they blend imperceptibly into legend and history. Through the deep pleasures they naturally excite, and the ease with which they are passed from one person to another, these stories invade the developing mind more readily than others, and they tend to converge to form the commonalities of human nature. Yeats in his 1900 essay on Shelley distinguished between the theoretician who seeks abstract truth and the naturalist-poet who celebrates detail. In the universe of the mind, Yeats said, no symbol tells all its meanings to any generation. Only by discovering the ancient symbols can the artist express meanings that cross generations and open the full abundance of nature.

We need not worry about the extravagances of visionary artists, so long as they reveal the deeper channels of their minds in a manner that gives meaning to our own. Each human mindscape is idiosyncratic and yet ultimately obedient to biological law. Like the forest of some newly discovered island, it possesses unique contours and previously undescribed forms of life, treasures to be valued for their own sake, but the genetic process that spawned them is the same as elsewhere. Continuity is essential for comprehension; the imagery chosen by the artist must draw on common experience and values, however tortuous the manner of presentation. Thus in 1919 the American modernist Joseph Stella created "Tree of My Life" to translate his own vaulting optimism into a physical paradise within the mind. Bright tropical plants and animals served as the symbols. He described his feelings that led to the painting:

> AND one clear morning in April I found myself in the midst of joyous singing and delicious scent—the singing and the scent of the birds and flowers ready to celebrate the baptism of my new art, the birds and the flowers already enjewelling the tender foliage of the new-born tree of my hopes.

We are in the fullest sense a biological species and will find little ultimate meaning apart from the remainder of life. The fiery circle of disciplines will be closed if science looks at the inward journey of the artist's mind, making art and culture objects of study in the biological mode, and if the artist and critic are informed of the workings of the mind and the natural world as illuminated by the scientific method. In principle at least, nothing can be denied to the humanities, nothing to science.

The Serpent

SCIENCE and the humanities, biology and culture, are bridged in a dramatic manner by the phenomenon of the serpent. The snake's image enters the conscious mind with ease during dreams and reverie, fabricated from symbols and bearing portents of magic. It appears without warning and departs abruptly, leaving behind not the perception of any real snake but the vague memory of a more powerful creature, the serpent, surrounded by a mist of fear and wonderment.

These qualities are dominant in one particular dream I have experienced often through my life, for reasons I will try to clarify later. I find myself in a locality that is wooded and aquatic, silent and drawn wholly in shades of gray. As I walk into this somber environment I am gripped by an alien feeling:

> The terrain before me is mysterious, on the rim of the unknown, at once calm and forbidding. I am required to be there but in the dream state cannot grasp the reasons. Suddenly the Serpent appears. It is not a snake of the ordinary kind, a literal reptile, but much more, a threatening presence with unusual powers. Its size is indeterminately large. While I watch its muscular coils slide into the water, beneath prop roots, and back onto the bank, protean in size and shape, armored, irresistible. The poisonous head radiates a cold, inhuman intelligence. The Serpent is somehow the spirit of that shadowed place and

> guardian of the passage into deeper reaches. I sense that if I could capture or control or even just evade it, a great change in the ambience would follow. The change cannot be defined immediately, but its anticipation stirs old and still unnamed emotions. The risk is also vaguely felt, like that emanating from a knife blade or high cliff. The Serpent is life-promising and life-threatening, seductive and treacherous. It now slips close to me, turning importunate, ready to strike. The dream ends uneasily, without clear resolution.

The snake and the serpent, flesh-and-blood reptile and demonic dream-image, reveal the complexity of our relation to nature and the fascination and beauty inherent in all forms of organisms. Even the deadliest and most repugnant creatures bring an endowment of magic to the human mind. Human beings have an innate fear of snakes or, more precisely, they have an innate propensity to learn such fear quickly and easily past the age of five. The images they build out of this peculiar mental set are both powerful and ambivalent, ranging from terror-stricken flight to the experience of power and male sexuality. As a consequence the serpent has become an important part of cultures around the world.

There is a principle of many ramifications to consider here, which extends well beyond the ordinary concerns of psychoanalytic reasoning about sexual symbols. Life of any kind is infinitely more interesting than almost any conceivable variety of inanimate matter. The latter is valued chiefly to the extent that it can be metabolized into live tissue, accidentally resembles it, or can be fashioned into a useful and properly animated artifact. No one in his right mind looks at a pile of dead leaves in preference to the tree from which they fell.

What is it exactly that binds us so closely to living things? The biologist will tell you that life is the self-replication of giant molecules from lesser chemical fragments, resulting in the assembly of complex organic structures, the transfer of large amounts of molecular information, ingestion, growth, movement of an outwardly purposeful nature, and the proliferation of closely similar organisms. The poet-in-biologist will add that life is an exceedingly improbable state, metastable, open to other systems, thus ephemeral—and worth any price to keep.

Certain organisms have still more to offer because of their special impact on mental development. I have suggested that the urge to affiliate with other forms of life is to some degree innate, hence deserves to be called biophilia. The evidence for the proposition is not strong in a formal scientific sense: the subject has not been studied enough in the scientific manner of hypothesis, deduction, and experimentation to let us be certain about it one way or the other. The biophilic tendency is nevertheless so clearly evinced in daily life and widely distributed as to deserve serious attention. It unfolds in the predictable fantasies and responses of individuals from early childhood onward. It cascades into repetitive patterns of culture across most or all societies, a consistency often noted in the literature of anthropology. These processes appear to be part of the programs of the brain. They are marked by the quickness and decisiveness with which we learn particular things about certain kinds of plants and animals. They are too consistent to be dismissed as the result of purely historical events working on a mental blank slate.

Perhaps the most bizarre of the biophilic traits is awe and veneration of the serpent. The dreams from which the dominant images arise are known to exist in all those societies where systematic studies have been conducted on mental life. At least 5 percent of the people at any given time remember experiencing them, while many more would probably do so if they recorded their waking impressions over a period of several months. The images described by urban New Yorkers are as detailed and emotional as those of Australian aboriginals and Zulus. In all cultures the serpents are prone to be mystically transfigured. The Hopi know Palulukon, the water serpent, a benevolent but frightening godlike being. The Kwakiutl fear the *sisiutl,* a kind of three-headed serpent with both human and reptile faces, whose appearance in dreams presages insanity or death. The Sharanahua of Peru summon reptile spirits by taking hallucinogenic drugs and stroking the severed tongues of snakes over their faces. They are rewarded with dreams of brightly colored boas, venemous snakes, and lakes teeming with caimans and anacondas. Around the world serpents and snakelike creatures are the dominant elements of dreams in which animals of any kind appear. Inspiring fear and veneration, they are recruited as the animate symbols of power and sex, totems, protagonists of myths, and gods.

These cultural manifestations may seem at first detached and mysterious, but there is a simple reality behind the ophidian archetype that lies within the experience of ordinary people. The mind is primed to react emotionally to the sight of snakes, not just to fear them but to be aroused and absorbed in their details, to weave stories about them. This distinctive predisposition played an important role in an unusual experience of my own. Let me tell a childhood story about an encounter with a large and memorable snake, a creature that actually existed.

I GREW UP in the Panhandle of northern Florida and the adjacent counties of Alabama, in circumstances that eventually turned me into a field biologist. Like most boys in that part of the country set loose to roam the woods, I enjoyed hunting and fishing and made no clear distinction between these activities and life at large. But I also cherished natural history for its own sake and decided very early to become a biologist. I had a secret ambition to find a Real Serpent, a snake so fabulously large or otherwise different that it would exceed the bounds of imagination.

Certain peculiarities in the environment encouraged this adolescent fantasy. It helped at the outset that I was an only child with indulgent parents, encouraged to develop my own interests and hobbies, however farfetched. In other words, I was spoiled. And except for issues pertaining to the literal claims of the King James Bible, our neighbors were equally tolerant of eccentric kids. They had to be: we all knew, even though we did not discuss openly, that certain families in the neighborhood kept very unusual children in their homes instead of placing them in institutions. It was a time in the South, about to come to a close, when family obligations and loyalties were unquestioned and spoken about mostly in oblique, ritual terms.

The physical surroundings inclined youngsters toward an awe of nature. That part of the country had been covered, four generations back, by a wilderness as formidable in some respects as the Amazon. Dense thickets of cabbage palmetto descended into meandering spring-fed streams and cypress sloughs. Carolina parakeets and ivory-billed woodpeckers flashed overhead in the sunlight, where wild turkeys and passenger pigeons could still be counted on as game. On soft spring nights after heavy rains a dozen varieties of frogs croaked, rasped, bonged, and trilled their love songs in mixed choruses. Much of the Gulf Coast fauna had been derived from species that spread north from the tropics over millions of years and adapted to the local, warm temperate conditions. Columns of miniature army ants, close replicas of the large marauders of South America, marched mostly unseen at night over the forest floor. *Nephila* spiders the size of saucers spun webs as wide as garage doors across the woodland clearings.

From the stagnant pools and knothole sinks, clouds of mosquitoes rose to afflict the early immigrants. They carried the Confederate plague, malaria and yellow fever, which periodically flared into epidemics and reduced the populations along the coastal lowlands. This natural check is one of the reasons the strip between Tampa and Pensacola remained sparsely settled for so long and why even today, long after the diseases have been eradicated, it is still the relatively natural "other Florida."

Snakes abounded. The Gulf Coast has a greater variety

and denser populations than almost any other place in the world, and they are frequently seen. Striped ribbon snakes hang in Gorgonlike clusters on branches at the edge of ponds and streams. Poisonous coral snakes root through the leaf litter, their bodies decorated with warning bands of red, yellow, and black. They are easily confused with their mimics, the scarlet kingsnakes, banded in a different sequence of red, black, and yellow. The simple rule recited by woodsmen is: "Red next to yellow will kill a fellow, Red next to black is a friend of Jack." Hognoses, harmless thick-bodied sluggards with upturned snouts, are characterized by an unsettling resemblance to venomous African gaboon vipers and a habit of swallowing toads live. Pygmy rattlesnakes two feet long contrast with diamondbacks seven feet long or more. Watersnakes are a herpetologist's medley told apart by size, color, and the arrangement of body scales. They comprise ten species of *Natrix, Seminatrix, Agkistrodon, Liodytes,* and *Farancia.*

Of course limits to the abundance and diversity exist. Because snakes feed on frogs, mice, fish, and other animals of similar size, they are necessarily scarcer than their prey. You can't just go out on a stroll and point to one individual after another. An hour's careful search will often turn up none at all. But I can testify from personal experience that on any given day you are ten times more likely to meet a snake in Florida than in Brazil or New Guinea.

There is something oddly appropriate about the abundance of snakes. Although the Gulf wilderness has been largely converted into macadam and farmland, and the sound of television and company jets is heard in the land, a remnant of the old rural culture remains, as if the population were still pitted against the savage and the unknown. "Push the forest back and fill the land" remains a common sentiment, the colonizer's ethic and tested biblical wisdom (the very same that turned the cedar groves of Lebanon into the fought-over desert they are today). The prominence of snakes lends symbolic support to this venerable belief.

In the back country during a century and half of settlement, the common experience of snakes was embroidered into the lore of serpents. Cut off a rattlesnake's head, one still hears, and it will live on until sundown. If a snake bites you, open the puncture wounds with a knife and wash them with kerosene to neutralize the poison (I never met anyone who claimed to have done that and survived). If you believe with all your heart in Jesus, you can hang rattlers and copperheads around your neck without fear. If one strikes you just the same, accept it as a sign from the Lord and find peace in whatever follows. The hognose snake, on the other hand, is always death in the shape of a slithery S. Those who get too close to one will have venom sprayed in their eyes and be blinded; the very breath from the snake's skin is lethal. This species is the beneficiary of its dreadful legend: I never heard of one being killed.

Deep in the woods live creatures of startling power. (*That* is what I most wanted to hear.) Among them is the hoop snake. Skeptics, who used to be found hunkered down in a row along the county courthouse guardrail on a Saturday morning, say it is only mythical; on the other hand it might be the familiar coachwhip racer turned vicious by special circumstances. Thus transformed, it puts its tail in its mouth and rolls down hills at great speed to attack its terrified victims. Then there were reports of the occasional true monsters: a giant snake believed to live in a certain swamp (used to be there anyway, even if no one's seen it in recent years); a twelve-foot diamondback rattler a farmer killed on the edge of town a few years back; some unclassifiable prodigy recently glimpsed as it sunned itself along the river's edge.

It is a wonderful thing to grow up in southern towns where animal fables are taken half seriously, breathing into the adolescent mind a sense of the unknown and the possibility that something extraordinary might be found within a day's walk of where you live. No such magic exists in the environs of Schenectady, Liverpool, and Darmstadt, and for all children dwelling in such places where the options have finally been closed, I feel a twinge of sadness. I found my way out of Mobile, Pensacola, and Brewton to explore the surrounding woods and swamps in a languorous mood. I formed the habit of quietude and concentration into which I still pass my mind during field excursions, having learned to summon the old emotions as part of the naturalist's technique.

Some of these feelings must have been shared by my friends. In the mid-1940s during the hot season between spring football practice and the regular schedule of games in the fall, working on highway cleanup gangs and poking around outdoors were about all we had to do. But there was some difference: I was hunting snakes with a passionate intensity. On the Brewton High School football team of 1944 – 45 most of the players had nicknames, leaning toward the infantilisms and initials favored by southerners: Bubba Joe, Flip, A. J., Sonny, Shoe, Jimbo, Junior, Snooker, Skeeter. As the underweight third-string left end, allowed to play only in the fourth quarter when the foe had been crushed beyond any hope of recovery, mine was Snake. And while of this measure of masculine acceptance I was inordinately proud, my main hopes and energies had been invested elsewhere. There are an incredible forty species of snakes native to that region, and I managed to capture almost all of them.

One kind became a special target just because it was so elusive: the glossy watersnake *Natrix rigida.* The adults lay on the bottom of shallow ponds well away from the shore and pointed their heads out of the alga-green water in order to breathe and scan the surface in all directions. I waded out toward them very carefully, avoiding the side-to-side movements to which snakes are most alert. I needed to get within three or four feet in order to manage a diving tackle, but before I could cover the distance they always pulled their heads under and slipped silently away into the opaque depths. I finally solved the problem with the aid of the town's leading slingshot artist, a taciturn loner my age, proud and quick to anger, the sort of boy who in an earlier time might have distinguished himself at Antietam or Shiloh. Aiming pebbles at the heads of the snakes, he was able to stun several long enough for me to grab them underwater. After recovering, the captives were kept for a while in homemade cages in our backyard, where they thrived on live minnows placed in dishes of water.

Once, deep in a swamp miles from home, half lost and not caring, I glimpsed an unfamiliar brightly colored snake disappearing down a crayfish burrow. I sprinted to the spot, thrust my hand after it and felt around blindly. Too late: the snake had squirmed out of reach into the lower chambers. Only later did I think about the possibilities. Suppose I had succeeded and the snake was poisonous? My reckless enthusiasm did catch up with me on another occasion when I miscalculated the reach of a pygmy rattlesnake, which struck out faster than I thought possible and hit me with startling authority on the left index finger. Because of the small size of the reptile, the only result was a temporarily swollen arm and a fingertip that still grows a bit numb at the onset of cold weather.

But I digress. I found my Serpent on a still July morning in the swamp fed by the artesian wells of Brewton, while working toward higher ground along the course of a weed-choked stream. Without warning a very large snake crashed away from under my feet and plunged into the water. Its movement startled me even more than it would have in other circumstances, because I had grown accustomed through the day to modestly proportioned frogs and turtles silently tensed on mudbanks and logs. This snake was more nearly my size as well as violent and noisy—a colleague, so to speak. It sped with wide body undulations to the center of the shallow watercourse and came to rest on a sandy riffle. It was not quite the monster I had envisioned but nevertheless unusual, a water moccasin *(Agkistrodon piscivorus),* one of the poisonous pit vipers, more than five feet long with a body as thick as my arm and a head the size of a fist. It was the largest snake I had ever seen in the wild. I later calculated it to be just under the published size record for the species. The snake now lay quietly in the shallow clear water completely open to view, its body stretched along the fringing weeds, its head pointed back at an oblique angle to watch my approach. Moccasins are like that. They don't always keep going until they are out of sight, in the manner of ordinary watersnakes. Although no emotion can be read in the frozen half-smile and staring yellow cat's eyes, their reactions and posture make them seem insolent, as if they see their power reflected in the caution of human beings and other sizable enemies.

I moved through the snake handler's routine: pressed the snake stick across the body in back of the head, rolled it forward to pin the head securely, brought one hand around to grasp the neck just behind the swelling masseteric muscles, dropped the stick to seize the body midway back with the other hand, and lifted the entire animal clear of the water. The technique almost always works. The moccasin, however, reacted in a way that took me by surprise and put my life in immediate danger. Throwing its heavy body into convulsions, it twisted its head and neck slightly forward through my gripped fingers, stretched its mouth wide open to unfold the inch-long fangs and expose the dead-white inner lining in the intimidating "cottonmouth" display. A fetid musk from its anal glands filled the air. At that moment the morning heat became more noticeable, the episode turned manifestly frivolous, and at last I wondered why I should be in that place alone. Who would find me? The snake began to turn its head far enough to clamp its jaws on my hand. I was not very strong for my age and I was losing control. Without thinking I heaved the giant out into the brush and it thrashed frantically away, this time until it was out of sight and we were rid of each other.

I sat down and let the adrenaline race my heart and bring tremors to my hand. How could I have been so stupid? What is there in snakes anyway that makes them so repellent and fascinating? The answer in retrospect is deceptively simple:

their ability to remain hidden, the power in their sinuous limbless bodies, and the threat from venom injected hypodermically through sharp hollow teeth. It pays in elementary survival to be interested in snakes and to respond emotionally to their generalized image, to go beyond ordinary caution and fear. The rule built into the brain in the form of a learning bias is: become alert quickly to any object with the serpentine gestalt. *Overlearn* this particular response in order to keep safe.

Other primates have evolved similar rules. When guenons and vervets, the common monkeys of the African forest, see a python, cobra, or puff adder, they emit a distinctive chuttering call that rouses other members in the group. (Different calls are used to designate eagles and leopards.) Some of the adults then follow the intruding snake at a safe distance until it leaves the area. The monkeys in effect broadcast a dangerous-snake alert, which serves to protect the entire group and not solely the individual who encountered the danger. The most remarkable fact is that the alarm is evoked most strongly by the kinds of snakes that can harm them. Somehow, apparently through the routes of instinct, the guenons and vervets have become competent herpetologists.

The idea that snake aversion on the part of man's relatives can be an inborn trait is supported by other studies on rhesus macaques, the large brown monkeys of India and surrounding Asian countries. When adults see a snake of any kind, they react with the generalized fear response of their species. They variously back off and stare (or turn away), crouch, shield their faces, bark, screech, and twist their faces into the fear grimace, in which the lips are retracted, the teeth are bared, and the ears are flattened against the head. Monkeys raised in the laboratory without previous exposure to snakes show the same response to them as those brought in from the wild, although in weaker form. During control experiments designed to test the specificity of the response, the rhesus failed to react to other, nonsinuous objects placed in their cages. It is the form of the snake and perhaps also its distinctive movements that contain the key stimuli to which the monkeys are innately tuned.

Grant for the moment that snake aversion does have a hereditary basis in at least some kinds of nonhuman primates. The possibility that immediately follows is that the trait evolved by natural selection. In other words, individuals who respond leave more offspring than those who do not, and as a result the propensity to learn fear quickly spreads through the population—or, if it was already present, is maintained there at a high level.

How can biologists test such a proposition about the origin of behavior? They turn natural history upside down. They search for species historically free of forces in the environment believed to favor the evolutionary change, to see if in fact the organisms do *not* possess the trait. The lemurs, primitive relatives of the monkeys, offer such an inverted opportunity. They are indigenous inhabitants of Madagascar, where no large or poisonous snakes exist to threaten them. Sure enough, lemurs presented with snakes in captivity fail to display anything resembling the automatic fear responses of the African and Asian monkeys. Is this adequate proof? In the chaste idiom of scientific discourse, we are permitted to conclude only that the evidence is consistent with the proposal. Neither this nor any comparable hypothesis can be settled by a single case. Only further examples can raise confidence in it to a level beyond the reach of determined skeptics.

Another line of evidence comes from studies of the chimpanzee, a species thought to have shared a common ancestor with prehumans as recently as five million years ago. Chimps raised in the laboratory become apprehensive in the presence of snakes, even if they have had no previous experience. They back off to a safe distance and follow the intruder with a fixed stare while alerting companions with the *Wah!* warning call. More important, the response becomes gradually more marked during adolescence.

This last quality is especially interesting because human beings pass through approximately the same developmental sequence. Children under five years of age feel no special anxiety over snakes, but later they grow increasingly wary. Just one or two mildly bad experiences, such as a garter snake seen writhing away in the grass, a playmate thrusting a rubber model at them, or a counselor telling scary stories at the campfire, can make children deeply and permanently fearful. The pattern is unusual if not unique in the ontogeny of human behavior. Other common fears, notably of the dark, strangers, and loud noises, start to wane after seven years of age. In contrast, the tendency to avoid snakes grows stronger with time. It is possible to turn the mind in the opposite direction, to learn to handle snakes without apprehension or even to like them in some special way, as I did—but the adaptation takes a special effort and is usually a little forced and self-conscious. The special sensitivity will just as likely lead to full-blown ophidiophobia, the pathological extreme in which the mere appearance of a snake brings on a feeling of panic, cold sweat, and waves of nausea. I have witnessed these events:

> At a campsite in Alabama, on a Sunday afternoon, a four-foot-long black racer glided out from the woods across the clearing and headed for the high grass along a nearby stream. Children shouted and pointed. A middle-aged woman screamed and collapsed to the ground sobbing. Her husband dashed to his pickup truck to get a shotgun. But black racers are among the fastest snakes in the world, and this one made it safely to cover. The onlookers probably did not know that the species is nonvenomous and harmless to any creature larger than a cotton rat.
>
> Halfway around the world, at the village of Ebabaang in New Guinea, I heard shouting and saw people running down a path. When I caught up with them they had formed a circle around a small brown snake that was

essing leisurely across the front yard of a house. I pinned the snake and carried it off to be preserved in alcohol for the museum collections at Harvard. This seeming act of daring earned either the admiration or the suspicion of my hosts — I couldn't be sure which. The next day children followed me around as I gathered insects in the nearby forest. One brought me an immense orb-weaving spider gripped in his fingers, its hairy legs waving and the evil-looking black fangs working up and down. I felt panicky and sick. It so happens that I suffer from mild arachnophobia. To each his own.

Why should serpents have such a strong influence during mental development? The direct and simple answer is that throughout the history of mankind a few kinds have been a major cause of sickness and death. Every continent except Antarctica has poisonous snakes. Over large stretches of Asia and Africa the known death rate from snake bite is 5 persons per 100,000 each year, or higher. The local record is held by a province in Burma, with 36.8 deaths per 100,000 a year. Australia has an exceptional abundance of deadly snakes, a majority of which are relatives of the cobras. Among them the tiger snake is especially feared for its large size and tendency to strike without warning. In South and Central America live the bushmaster, fer-de-lance, and jaracara, among the largest and most aggressive of the pit vipers. With backs colored like rotting leaves and fangs long enough to pass through a human hand, they lie in ambush on the floor of the tropical forest for the small warm-blooded animals that form their major prey. Few people realize that a complex of dangerous snakes, the "true" vipers, are still relatively abundant throughout Europe. The common adder *Viperus berus* ranges to the Arctic Circle. The number of people bitten in such improbable places as Switzerland and Finland is still high enough, running into the hundreds annually, to keep outdoorsmen on a sort of yellow alert. Even Ireland, one of the few countries in the world lacking snakes altogether (thanks to the last Pleistocene glaciation and not Saint Patrick), has imported the key ophidian symbols and traditions from other European cultures and preserved the fear of serpents in art and literature.

HERE, THEN, is the sequence by which the agents of nature appear to have been translated into the symbols of culture. For hundreds of thousands of years, time enough for the appropriate genetic changes to occur in the brain, poisonous snakes have been a significant source of injury and death to human beings. The response to the threat is not simply to avoid it, in the way that certain berries are recognized as poisonous through a process of trial and error. People also display the mixture of apprehension and morbid fascination characterizing the nonhuman primates. They inherit a strong tendency to acquire the aversion during early childhood and to add to it progressively, like our closest phylogenetic relatives, the chimpanzees. The mind then adds a great deal more that is distinctively human. It feeds upon the emotions to enrich culture. The tendency of the serpent to appear suddenly in dreams, its sinuous form, and its power and mystery are the natural ingredients of myth and religion.

Consider how sensation and emotional states are elaborated into stories during dreams. The dreamer hears a distant thunderclap and changes an ongoing episode to end with the slamming of a door. He feels a general anxiety and is transported to a schoolhouse corridor, where he searches for a classroom he does not know in order to take an examination for which he is unprepared. As the sleeping brain enters its regular dream periods, marked by rapid eye movement beneath closed eyelids, giant fibers in the lower brainstem fire upward into the cortex. The awakened mind responds by retrieving memories and fabricating stories around the sources of physical and emotional discomfort. It hastens to recreate the elements of past real experience, often in a jumbled and antic form. And from time to time the serpent appears as the embodiment of one or more of these feelings. The direct and literal fear of snakes is foremost among them, but the dream-image can also be summoned by sexual desire, a craving for dominance and power, and the apprehension of violent death.

We need not turn to Freudian theory in order to explain our special relationship to snakes. The serpent did not originate as the vehicle of dreams and symbols. The relation appears to be precisely the other way around and correspondingly easier to study and understand. Humanity's concrete experience with poisonous snakes gave rise to the Freudian phenomena after it was assimilated by genetic evolution into the brain's structure. The mind has to create symbols and fantasies from something. It leans toward the most powerful preexistent images or at least follows the learning rules that create the images, including that of the serpent. For most of this century, perhaps overly enchanted by psychoanalysis, we have confused the dream with the reality and its psychic effect with the ultimate cause rooted in nature.

Among prescientific people, whose dreams are conduits to the spirit world and snakes a part of ordinary experience, the serpent has played a central role in the building of culture. There are magic incantations for simple protection as in the hymns of the Atharva Veda:

> With my eye do I slay thy eye, with poison do I slay thy poison. O Serpent, die, do not live; back upon thee shall thy poison turn.

"Indra slew thy first ancestors, O Serpent," the chant continues, "and since they are crushed, what strength forsooth can be theirs?" And so the power can be controlled and even diverted to human use through iatromancy and the casting of magic spells. Two serpents entwine the caduceus, which was first the winged staff of Mercury as messenger of the gods, then the safe-conduct pass of ambassadors and heralds, and finally the universal emblem of the medical profession.

Balaji Mundkur has shown how the inborn awe of snakes matured into rich productions of art and religion around the world. Serpentine forms wind across stone carvings from paleolithic Europe and are scratched into mammoth teeth found in Siberia. They are the emblems of power and ceremony for the shamans of the Kwakiutl, the Siberian Yakut and Yenisei Ostyak, and many of the tribes of Australian aboriginals. Stylized snakes have often served as the talismans of the gods and spirits who bestow fertility: Ashtoreth of the Canaanites, the demons Fu-Hsi and Nu-kua of the Han Chinese, and the powerful goddesses Mudammā and Manasā of Hindu India. The ancient Egyptians venerated at least thirteen ophidian deities ministering to various combinations of health, fecundity, and vegetation. Prominent among them was the triple-headed giant Nehebkau who traveled widely to inspect every part of the river kingdom. Amulets in gold inscribed with the sign of a cobra god were placed in the wrappings of Tutankhamen's mummy. Even the scorpion goddess Selket bore the title "mother of serpents." Like her offspring she prevailed simultaneously as a source of evil, power, and goodness.

The Aztec pantheon was a phantasmagoria of monstrous forms among whom serpents were given pride of place. The calendrical symbols included the ophidian *olin nahui* and *cipactli,* the earth crocodile that possessed a forked tongue and rattlesnake's tail. The rain god Tlaloc consisted in part of two coiled rattlesnakes whose heads met to form the god's upper lip. *Coatl,* serpent, is the dominant phrase in the names of the divinities. Coatlicue was a threatening chimera of snake and human parts, Cihuacoatl the goddess of childbirth and mother of the human race, and Xiuhcoatl the fire serpent over whose body fire was rekindled every fifty-two years to mark a major division in the religious calendar. Quetzalcoatl, the plumed serpent with a human head, reigned as god of the morning and evening star and thus of death and resurrection. As inventor of the calendar, deity of books and learning, and patron of the priesthood, he was revered in the schools where nobles and priests were taught. His reported departure over the eastern horizon upon a raft of snakes must have been the occasion of consternation for the intellectuals of the day, something like the folding of the Guggenheim Foundation.

A contradiction of ophidian images was a feature of Greek religion as well. Among the early forms of Zeus was the serpent Meilikhios, god of love, gentle and responsive to supplication, and god of vengeance, whose sacrifice was a holocaust offered at night. Another great serpent protected the lustral waters at the spring of Ares. He coexisted with the Erinyes, demons of the underworld so horrible they could not be pictured in early mythology. They were given the form of serpents when brought to stage by Euripides in the *Iphigeneia in Tauris:* "Dost see her, her the Hades-snake who gapes / To slay me, with dread vipers, open-mouthed?"

Slyness, deception, malevolence, betrayal, the implicit threat of a forked tongue flicking in and out of the masklike head, all qualities tinged with miraculous powers to heal and guide, forecast and empower, became the serpent's prevailing image in western cultures. The serpent in the Garden of Eden, appearing as in a dream to serve as Judaism's evil Prometheus, gave humankind knowledge of good and evil and with it the burden of original sin, for which God repaid in kind:

> I will put enmity between you and the woman,
> between your brood and hers.
> They shall strike at your head,
> and you shall strike at their heel.

TO SUMMARIZE the relation between man and snake: life gathers human meaning to become part of us. Culture transforms the snake into the serpent, a far more potent creation than the literal reptile. Culture in turn is a product of the mind, which can be interpreted as an image-making machine that recreates the outside world through symbols arranged into maps and stories. But the mind does not have an instant capacity to grasp reality in its full chaotic richness; nor does the body last long enough for the brain to process information piece by piece like an all-purpose computer. Rather, consciousness races ahead to master certain kinds of information with enough efficiency to survive. It submits to a few biases easily while automatically avoiding others. A great deal of evidence has accumulated in genetics and physiology to show that the controlling devices are biological in nature, built into the sensory apparatus and brain by particularities in cellular architecture.

The combined biases are what we call human nature. The central tendencies, exemplified so strikingly in fear and veneration of the serpent, are the wellsprings of culture. Hence simple perceptions yield an unending abundance of images with special meaning while remaining true to the forces of natural selection that created them.

How could it be otherwise? The brain evolved into its present form over a period of about two million years, from the time of *Homo habilis* to the late stone age of *Homo sapiens,* during which people existed in hunter-gatherer bands in intimate contact with the natural environment. Snakes mattered. The smell of water, the hum of a bee, the directional bend of a plant stalk mattered. The naturalist's trance was adaptive: the glimpse of one small animal hidden in the grass could make the difference between eating and going hungry in the evening. And a sweet sense of horror, the shivery fascination with monsters and creeping forms that so delights us today even in the sterile hearts of the cities, could see you through to the next morning. Organisms are the natural stuff of metaphor and ritual. Although the evidence is far from all in, the brain appears to have kept its old capacities, its channeled quickness. We stay alert and alive in the vanished forests of the world.

The Right Place

THE NATURALIST is a civilized hunter. He goes alone into a field or woodland and closes his mind to everything but that time and place, so that life around him presses in on all the senses and small details grow in significance. He begins the scanning search for which cognition was engineered. His mind becomes unfocused, it focuses on everything, no longer directed toward any ordinary task or social pleasantry. He measures the antic darting of midges in a conical mating swarm, the slant of sunlight by which they are best seen, the precise molding of mosses and lichens on the tree trunk on which they spasmodically alight. His eye travels up the trunk to the first branch and out to a spray of twigs and leaves and back, searching for some irregularity of shape or movement of a few millimeters that might betray an animal in hiding. He listens for any sound that breaks the lengthy spells of silence. From time to time he translates his running impressions of the smell of soil and vegetation into rational thought: the ancient olfactory brain speaks to the modern cortex. The hunter-in-naturalist knows that he does not know what is going to happen. He is required, as Ortega y Gasset expressed it, to prepare an attention of a different and superior kind, "an attention that does not consist in riveting itself to the presumed but consists precisely in not presuming anything and avoiding inattentiveness."

Every practicing naturalist has favorite stories to tell about the rewards of chance in the field. I once went out with

Jesse Nichols, a professional animal collector, to a woodlot in central Alabama late at night in a cold rain to look for frogs and salamanders. I had been to the site several times before on sunny days and seen nothing. That night, as soon as we walked into the woods, we found a teeming population of a pygmy salamander in the genus *Desmognathus,* recently described by zoologists as a new species. The delicately built amphibians, which resembled shiny popeyed lizards, were climbing up grass and low bushes. They jumped agilely from branch to branch in search of prey. It had been our good fortune to encounter them under the most favorable environmental conditions at the height of their activity, and it occurred to us that as a result we had made a worthwhile discovery about desmognaths in general. This class of salamanders usually lives at the edge of water or concealed in litter and soil. We now knew that one of the species is also partly arboreal and behaves a bit like tree frogs. It follows that desmognath salamanders as a whole are more ecologically diverse than originally believed. They have undergone a moderate evolutionary expansion near the center of their range, a broad area in the southeastern United States that includes Alabama. We discussed these weighty matters as we shivered in the rain, plucking enough pygmy salamanders off bushes to give to museums around the country.

Field research consists of hard physical work broken by moments of happy surprise. In his autobiography William Mann, until 1958 Director of the National Zoological Park in Washington but by training an entomologist, tells of a trip he made as a young man into the Sierra de Trinidad of central Cuba. When he lifted a rock to see what animals were hiding underneath (there are always animals of some kind, usually very small, under every rock), it split down the middle to expose a half-teaspoonful of metallic-green ants living in a small cavity deep inside. Mann went on to name this remarkable creature *Macromischa wheeleri,* in honor of William Morton Wheeler, his major professor at Harvard and the reigning world authority on ants. Thirty-six years later, with his discovery a romantic image in my head, I was climbing a steep slope in the same mountains, another young man at the start of a career in entomology. I had begun an ant-hill odyssey around the world remarkably similar to Mann's. A rock I grabbed for support split in my hand, exposing a half-teaspoonful of the same glittering green species. I accepted the event as one of the rites of passage.

Within hours on the same hillside I had another piece of luck. I obtained a live adult of the rare giant anole lizard, *Chamaeleolis chamaeleontides,* found only in Cuba and previously something of a mystery to biologists. It belongs to a group of species sometimes called the false chameleons because of an ability, shared with the true chameleons of Africa, to change skin color according to background and mood. The foot-long lizard also had a naturally wrinkled skin and a tired-looking expression, and I named my specimen Methuselah. During the remainder of my travels in Cuba and Mexico that summer of 1953 and after we came back to Cambridge in the fall, Methuselah spent much of his time riding on my shoulders. By watching him almost daily over a six-month's period as I fed him live mealworms and other insects, I came to realize that *Chamaeleolis* closely resembles the African chameleons in behavior as well as appearance. Both hunt with a slowness and deliberateness unusual for lizards, swivel their partly fused eyelids around to change the field of vision, and capture prey by flicking out long sticky tongues at nearly invisible speed. The similarity provided one more textbook example of evolutionary convergence between separate lines of animals that originated in the Old and New Worlds, in this case Africa as opposed to Cuba. The facts (which I published in a short technical article) were less than earth-shaking, but solid and satisfying—at least they will outlive Methuselah and me.

INSOFAR as organisms have been scrutinized, the naturalist can place them: their linkage in the ecosystem, life cycle, behavior, genetics, evolutionary history, physiology, and from all this information something of their general significance by whatever philosophy guided the naturalist to his life's pursuit in the first place. He is conducting a hunt in another mode, not for the animal's body but for discoveries, new information that will become part of the permanent record about the species viewed as an enduring entity. The pursuit is peculiarly satisfying because it enters that part of the real world, largely unrecognized, where humanity evolved during most of its two-million-year history. The vivifying eye of the naturalist is the orderly response to the original human environment.

What was that environment? To answer the question, we must turn natural history partly into an exercise in aesthetic judgment. The more habitats I have explored, the more I have felt that certain common features subliminally attract and hold my attention. Is it unreasonable to suppose that the human mind is primed to respond most strongly to some narrowly defined qualities that had the greatest impact on survival in the past? I am not suggesting the existence of an instinct. There is no evidence of a hereditary program hardwired into the brain. We learn most of what we know, but some things are learned much more quickly and easily than others. The hypothesis of biased learning is at least worth examining, and the logical point of departure is a pair of derived questions. What was the prevailing original habitat in which the brain evolved? Where would people go if given a completely free choice?

The whole matter may seem imponderable at first, but a workable approach can be found in this generalization from ecology: the crucial first step to survival in all organisms is habitat selection. If you get to the right place, everything else is likely to be easier. Prey become familiar and vulnerable, shelters can be put together quickly, and predators are tricked and beaten consistently. A great many of the complex structures in the sense organs and brain that characterize each species serve the primary function of habitat selection. They determine the sounds, sights, and smells individuals receive and the sequence of responses these stimuli evoke.

Following inborn rules of behavior, animals turn to the special routes and crannies for which the remainder of their anatomy and physiology is particularly well suited. A few make crucial choices in the first few minutes of their lives. In what may be the ultimate case among mammals, the newborn kangaroo travels over its mother's belly from her genital opening to the nipples located deep in the pouch. Because it is completely blind, the peanut-sized creature must rely on a precise instinctual reading of the odor and feel of every centimeter of fur. In order to duplicate an equal feat of orientation, a human infant would have to emerge unaided from the womb, crawl down onto the carpet, make its way directly through the house to the nursery and into the crib, seize a bottle and start feeding.

So precise is habitat selection by many animals that closely related species can often be told apart more quickly by where they are found than by any obvious physical trait. The North American flycatchers, for example, are relatively small, inconspicuously colored birds that flit in and out of trees to snatch insects from the air. Only an expert can separate the species readily by outward appearance alone, but even a beginner can make a reliable identification if the habitat is added in. The alder flycatcher lives primarily in swamps and wet thickets, while each of the other species chooses a special combination of sites from among coniferous forests, cold bogs, farmland, and open mixed woodland.

Even more instructive is the case of the prairie deer mouse of the central United States. Wild populations remain strictly in open terrain, avoiding all kinds of forests, even those with grassy floors. When biologists raised individuals in outdoor enclosures simulating the principal natural environments, they found that the orientation is inborn, although it can be additionally reinforced by early exposure to open places. They were also able to breed the trait out of captive deer mouse populations in less than twenty generations, so that afterward individual mice were just as likely to enter woods as fields.

Salamanders, frogs, and insects make finer discriminations appropriate to their smaller size. They settle on precisely defined sites beneath stones or on vegetation that offer the optimal combination of moisture, light, and temperature for their species. Even colon bacteria swim skillfully to the position in a drop of water where nutrients are most concentrated—but in a decidedly peculiar way. They move by spinning the whiplike flagellum at the end of the body like a ship's propeller. If the effort takes a particular bacterium from a higher to a lower concentration, in other words away from the nutrients, the organism changes course by reversing the spin, forcing the filaments of its flagellum to fly apart. This action makes it tumble through the water. When the tumbling stops, the filaments come together again, allowing the bacterium to swim in a new direction. Eventually, by trial and error, it reaches a zone of concentration high enough to let it feed. Microbiologists have succeeded in locating the genes and sensitive proteins that guide this simplest of all known orientation devices. They have identified mutations that change the structure of the controlling molecules and hence the direction in which bacteria swim. An important test of evolutionary theory has been passed: it is possible to alter an organism so that it automatically chooses the wrong habitat and condemns itself to death.

The question of interest is the preferred habitat of human beings. It is often said that *Homo sapiens* is the one species that can live anywhere—on top of ice floes, inside caves, under the sea, in space, anywhere—but this is just a half truth. People must jigger their environment constantly in order to keep it within a narrow range of atmospheric conditions. And once they have managed to rise above the level of bare subsistence, they invest large amounts of time to improve the appearance of their immediate surroundings. Their aim is to make the habitat more "livable" according to what are usually called aesthetic criteria.

With aesthetics we return to the central issue of biophilia. It is interesting to inquire about the prevalent direction of this vector in cultural evolution, in other words the ideal toward which human beings unconsciously strive no less relentlessly than flycatchers and deer mice. For if animals choose habitats by orientation devices and prepared learning built in during generations of natural selection, it is possible that people do the same. If certain human feelings are innate, they might not be easily expressed in rational language. A more promising approach is to explore the nature of the environment in which the brain evolved. The logical hypothesis I raised earlier can then be more precisely expressed. It is that certain key features of the ancient physical habitat match the choices made by modern human beings when they have a say in the matter.

The archeological evidence seems clear on the question of the original environment. For most of two million years human beings lived on the savannas of Africa, and subsequently those of Europe and Asia, vast, parklike grasslands dotted by groves and scattered trees. They appear to have avoided the equatorial rain forests on one side and the deserts on the other. There was nothing foreordained about this choice. The two extreme habitats have no special qualities that deny them to primates. Most monkeys and apes flourish in the rain forest, and two species, the hamadryas baboon and gelada, are specialized for life in the relatively barren grasslands and semideserts of Africa. The prehistoric species of *Homo* can be viewed both as the progenitors of modern human beings and as one more product among many within the great primate radiation of the Old World. In the latter role they belong to the minority of species that hit upon an intermediate topography, the tropical savanna. Most students of early human evolution agree that the bipedal locomotion and free-swinging arms fitted these ancestral forms very well to the open land, where they were able to exploit an abundance of fruits, tubers, and game.

THE BODY — YES. But is the *mind* predisposed to life on the savanna, such that beauty in some fashion can be said to lie in the genes of the beholder? Three scientists, Gordon

Orians, Yi-Fu Tuan, and the late René Dubos, have independently suggested that this is indeed the case. They point out that people work hard to create a savanna-like environment in such improbable sites as formal gardens, cemeteries, and suburban shopping malls, hungering for open spaces but not a barren landscape, some amount of order in the surrounding vegetation but less than geometric perfection. Orians in particular has elaborated the idea according to modern evolutionary theory and added a small but suggestive body of supporting evidence. According to his formulation, the ancestral environment contained three key features.

First, the savanna by itself, with nothing more added, offered an abundance of animal and plant food to which the omnivorous hominids were well adapted, as well as the clear view needed to detect animals and rival bands at long distances. Second, some topographic relief was desirable. Cliffs, hillocks, and ridges were the vantage points from which to make a still more distant surveillance, while their overhangs and caves served as natural shelters at night. During longer marches, the scattered clumps of trees provided auxiliary retreats sheltering bodies of drinking water. Finally, lakes and rivers offered fish, mollusks, and new kinds of edible plants. Because few natural enemies of man can cross deep water, the shorelines became natural perimeters of defense.

Put these three elements together: it seems that whenever people are given a free choice, they move to open tree-studded land on prominences overlooking water. This worldwide tendency is no longer dictated by the hard necessities of hunter-gatherer life. It has become largely aesthetic, a spur to art and landscaping. Those who exercise the greatest degree of free choice, the rich and powerful, congregate on high land above lakes and rivers and along ocean bluffs. On such sites they build palaces, villas, temples, and corporate retreats. Psychologists have noticed that people entering unfamiliar places tend to move toward towers and other large objects breaking the skyline. Given leisure time, they stroll along shores and river banks. They look along the water and up, to the hills beyond or to high buildings, expecting to see the sacred and beautiful places, the sites of historic events, now the seats of government, museums, or the homes of important personages. And they often do, in such landmarks as the Zähringen-Kyburg fortress of Thun, the Belvedere palace of Vienna, the cathedral of Saint Etienne, the chateau of Angers, and the Potala, and among the more imposing sites from past eras, Thingvellir, location of the ancient parliament of Iceland, the Parthenon, and the great plaza at Tenochtitlán.

The most revealing manifestation of the triple criterion occurs in the principles of landscape design. When people are confined to crowded cities or featureless land, they go to considerable lengths to recreate an intermediate terrain, something that can tentatively be called the savanna gestalt. At Pompeii the Romans built gardens next to almost every inn, restaurant, and private residence, most possessing the same basic elements: artfully spaced trees and shrubs, beds of herbs and flowers, pools and fountains, and domestic statuary. When the courtyards were too small to hold much of a garden, their owners painted attractive pictures of plants and animals on the enclosure walls—in open geometric assemblages. Japanese gardens, dating from the Heian period of the ninth to twelfth centuries (and hence ultimately Chinese in origin), similarly emphasize the orderly arrangement of trees and shrubs, open space, and streams and ponds. The trees have been continuously bred and pruned to resemble those of the tropical savanna in height and crown shape. The dimensions are so close as to make it seem that some unconscious force has been at work to turn Asiatic pines and other northern species into African acacias.

I will grant at once the strangeness of the comparison and the possibility that the convergence is merely a large coincidence. It is also true that individuals often yearn to retain the dominant and sometimes peculiar qualities of the environment in which they were raised. But entertain for a while longer the idea that the landscape architects and gardeners, and we who enjoy their creations without special instruction or persuasion, are responding to a deep genetic memory of mankind's optimal environment. That given a completely free choice, people gravitate statistically toward a savanna-like environment. The theory accommodates a great many seemingly disconnected facts from other parts of the world.

Far away, on the western frontier of the United States, explorers were given a brief opportunity to select the landscape to which their hearts led them. In their journals and memoirs they made clear the habitat they most valued. Not the dark forest, waiting to be cut back and replaced with a pastoral landscape of crops and hedges. Not the empty desert flats, good only if irrigated and planted in grass and trees. But the intermediate habitat already in place, a terrain that we ourselves can instantly appreciate: a savanna, rolling gold and green, dissected by a sharp tracery of streams and lake, with clean dry air and clouds dappling a blue sky. When Captain R. B. Marcy, on a United States government expedition through the southern plains in 1849, encountered the land around the headwaters of the Clear Fork of the Brazos River, he declared it to be "as beautiful a country for eight miles as I ever beheld."

> It was a perfectly level grassy glade, and covered with a growth of large mesquite trees at uniform distances, standing with great regularity, and presenting more the appearance of an immense peach orchard than a wilderness. The grass is of the short buffalo variety and as uniform and even as new mown meadow, and the soil is as rich, and very similar to that of the Red River Bottoms.

W. P. Parker, Marcy's companion, agreed: "The view was the most extensive and glowing in the sunset, the most striking that we had enjoyed during the whole trip, combining the grandeur of immense space—the plain extending to the horizon on every side from our point of view—with the beauty of the contrast between the golden carpet of buffalo grass and the pale green of the mesquite trees dotting its surface."

A note on botany: the trees are mesquite, a mimosaceous tree-shrub. The Brazos country is a passable replica of the tropical savanna with a dominant life form closely related to the African acacias, which are also members of the Mimosaceae. I have felt a similar attraction while traveling through the sawgrass and buttonbush flats of the Florida Everglades, the eucalyptus woodland of Queensland, and most compellingly the immense virgin savannas of South America.

Not long ago I joined a group of Brazilian scientists on a tour of the upland savanna, the *cerrado,* around the capital city of Brasilia. We went straight to one of the highest elevations as if following an unspoken command. We looked out across the rippled terrain of high grass, parkland, and forest enclaves and watched birds circling in the sky. We scanned the cumulus clouds that tower like high mountains above the plains during the wet season and found a gray curtain of rain descending into a valley behind some distant hills. We traced gallery forests, groves of trees that wind along the banks of the widely spaced streambeds. We studied Brasilia itself, now almost at the horizon, to admire the shining buildings and monuments that rise like well-spaced terraced cliffs and giant trees, and discussed the green belt and artificial lake that were designed and executed to make existence more livable—more human. It was, all agreed, very beautiful. Of such feelings Melville wrote, "Were Niagara but a cataract of sand, would you travel your thousand miles to see it?"

The practical-minded will argue that certain environments are just "nice" and there's an end to it. So why dilate on the obvious? The answer is that the obvious is usually profoundly significant. Some environments are indeed pleasant, for the same general reason that sugar is sweet, incest and cannibalism repulsive, and team sports exhilirating. Each response has its peculiar meaning rooted in the distant genetic past. To understand why we have one particular set of ingrained preferences, and not another, out of the vast number possible remains a central question in the study of man.

It might still be argued that people are just tracking ideal features of an environment sought out by other creatures as well. If that were true, the whole issue would be trivialized. If the most general properties of human nature are shared with lower organisms in a manner similar to eating and elimination, they could be studied more efficiently in simple animals such as squirrels and bobolinks. But such is not the case. Although the rules of sexual choice, diet selection, and social behavior are to some extent shared with a few other species, the overall pattern is particular to *Homo sapiens.* Not only symbolization and language, but also most of the basic cognitive specializations are unique. Among them appears to be biophilia, which is richly structured and quite irrational, in conformity with a primate genetic history played out in the warm climates of the Old World. Arcturian zoologists visiting this planet could make no sense of our morality and art until they reconstructed our genetic history—nor can we.

THERE IS ANOTHER way to measure the strength of human biophilia. Visualize a beautiful and peaceful world, where the horizon is rimmed by snowy peaks reaching into a perfect sky. In the central valley, waterfalls tumble down the faces of steep cliffs into a crystalline lake. On the crest of the terminal bluff sits a house containing food and every technological convenience. Artisans have worked across the terrain below to create a replica of one of Earth's landscape treasures, perhaps a formal garden from late eighteenth-century England, or the Garden of the Golden Pavilion at Kyoto, marked by an exquisite balance of water, copse, and trail. The setting is the most visually pleasing that human imagination can devise. Except for one thing—it contains no life whatever. This world has always been dead. The vegetation of the garden is artificial, shaped from plastic and colored by master craftsmen down to the last blade and stem. Not a single microbe floats in the lake or lies dormant in the ground. The only sounds are the broken rhythms of the falling water and an occasional whisper of wind through the plastic trees.

Where are we? If the ultimate act of cruelty is to promise everything and withhold just the essentials, the locality is a department of hell. It is a tomb built on a lunar landscape with air and elaborate contrivances added. This is a world (and more than a theoretical possibility in the age of space travel) where people would find their sanity at risk. Without beauty and mystery beyond itself, the mind by definition is deprived of its bearings and will drift to simpler and cruder configurations. Artifacts are incomparably poorer than the life they are designed to mimic. They are only a mirror to our thoughts. To dwell on them exclusively is to fold inwardly over and over, losing detail at each translation, shrinking with each cycle, finally merging into the lifeless facade of which they are composed.

When exceptions occur, they are incomplete and temporary. A few people can escape for a time into a world consisting exclusively of themselves and their machines and exist there without noticeable loss, providing they have strong character and a clearly defined goal. When Cyril Smith began his career in metallurgy as an employee of the American Brass Company, he treated the fire and clangor of the foundries as an aesthetic experience:

> I still have vivid sensual memories of that time: The smell of burning lard oil. Streams of molten brass in the casting shop. Some of the last coke-fired pit furnaces in operation, and men drawing crucibles, skimming and pouring the metal. The magnificent row of rolling mills, all driven continuously by a Corliss engine with a huge flywheel and a shaft running the full length of the large shop. The dance and clangor of drop and screw presses . . . To this day a frequent dream is of wandering through complex assemblies of industrial buildings full of such machines, in search of something I never find.

But Smith was no satanic apprentice adapted to an artifactual world. He was attracted to the most swiftly changing and visually dramatic events, in other words to quasi-life, one might say ultimately back toward life itself. Even in his anxiety dreams he searched for new and undefined experiences of similar kind. When he expanded these themes in his autobiographical *A Search for Structure,* he compared the most attractive patterns of the physical world and technology to artistic representations of plants and animals. People react more quickly and fully to organisms than to machines. They will walk into nature, to explore, hunt, and garden, if given the chance. They prefer entities that are complicated, growing, and sufficiently unpredictable to be interesting. They are inclined to treat their most formidable contraptions as living things or at least to adorn them with eagles, floral friezes, and other emblems representative of the peculiar human perception of true life. The ultimate machine of the futurist's imagination is a self-replicating robot that is benignly independent of its creators, hence in key respects quasi-alive. Mechanophilia, the love of machines, is but a special case of biophilia.

These qualities should impart a certain reserve about man's destiny out there among the stars. Let me qualify that remark at once. As a scientist and hence professional optimist, I am inspired perhaps more than most by the exploration of space. Our knowledge and self-understanding have been greatly expanded by orbiting scanners, probes, soft landings; and the technical spinoff seems to have no limit. If we can stripmine the moon, sweep rare elements from a comet's tail, and change the atmosphere of Venus to resemble Earth's ("terraforming it" is the favored expression), we should not hesitate—so long as the practical and scientific benefits are commensurate to the costs.

But the actual colonization of space by human beings is another matter altogether. No one doubts that the venture has compelling virtues. It would vault the historic expansion of the species around the world out to the unlimited frontiers beyond the planet, feeding the best in the human spirit. It would blast surplus populations from the source of their (more important, *our*) problems. The pioneer of this dream,

Gerard O'Neill, and other experts including NASA engineers have explored the technical aspects of the project and are sure it can be done. The gigantic cylinders and toroids they envision are admirable in scope and ingenuity. The interiors will be lined with agricultural fields, parks, and lakes, already depicted persuasively in preliminary layout paintings. These visualizations clearly reflect the designers' unconscious concession to the pull of the primitive human environment. And therein lies the problem as I see it.

For tugging at the bottom of the minds of the planners is an awareness that the mental health of the colonists is as important as their physical well-being. The whole enterprise is afflicted by an unsolved problem of unknown magnitude: can the psychic thread of life on Earth be snapped without eventually fatal consequences? A stable ecosystem can probably be created from an eternal cycling of microorganisms and plants. But it would still be an island of minute dimensions desperately isolated from the home planet, and simpler and less diverse by orders of magnitude than the environment in which human beings evolved. The tedium in such a reduced world would be oppressive for highly trained people aware of the grandeur of the original biosphere.

Even more painful would be the responsibility for keeping the station alive. There is a fundamental difference between the projected mental life of space colonies and ordinary mental life on Earth. It is far more frightening to know that only expert human intervention prevents the whole world from collapsing than merely to know that human beings can destroy it if they try. The comparison is similar to maintaining a patient in intensive care as opposed to watching him walk down the street in good health. People cannot be expected to carry such a burden; they were not built to be godlike in this particular sense. So when we dream of human populations expanding through the solar system and beyond, I believe we dream too far.

The chief significance of the life-in-space debate is symbolic rather than practical. Space colonies are very far down on the list of public priorities and not likely to be undertaken for generations — being part of the agenda, as it were, of the twenty-first century. They are useful right now for what they reveal about the poverty of our self-knowledge. The audaciously destructive tendencies of our species run deep and are poorly understood. They are so difficult to probe and manage as to suggest an archaic biological origin. We run a risk if we continue to diagnose them as by-products of history and suppose that they can be erased with simple economic and political remedies. At the very least, the Sophoclean flaws of human nature cannot be avoided by an escape to the stars. If people perform so badly on Earth, how can they be expected to survive in the biologically reduced and more demanding conditions of space?

Surely we would be better advised to invest the money on the workings of the mind. We should pay more attention to the quality of our dependence on other organisms. The brain is prone to weave the mind from the evidences of life, not merely the minimal contact required to exist, but a luxuriance and excess spilling into virtually everything we do. People can grow up with the outward appearance of normality in an environment largely stripped of plants and animals, in the same way that passable looking monkeys can be raised in laboratory cages and cattle fattened in feeding bins. Asked if they were happy, these people would probably say yes. Yet something vitally important would be missing, not merely the knowledge and pleasure that can be imagined and might have been, but a wide array of experiences that the human brain is peculiarly equipped to receive. Of that much I feel certain, and I will offer it in the form of a practical recommendation: on Earth no less than in space, lawn grass, potted plants, caged parakeets, puppies, and rubber snakes are not enough.

The Conservation Ethic

WHEN VERY LITTLE is known about an important subject, the questions people raise are almost invariably ethical. Then as knowledge grows, they become more concerned with information and amoral, in other words more narrowly intellectual. Finally, as understanding becomes sufficiently complete, the questions turn ethical again. Environmentalism is now passing from the first to the second phase, and there is reason to hope that it will proceed directly on to the third.

The future of the conservation movement depends on such an advance in moral reasoning. Its maturation is linked to that of biology and a new hybrid field, bioethics, that deals with the many technological advances recently made possible by biology. Philosophers and scientists are applying a more formal analysis to such complex problems as the allocations of scarce organ transplants, heroic but extremely expensive efforts to prolong life, and the possible use of genetic engineering to alter human heredity. They have only begun to consider the relationships between human beings and organisms with the same rigor. It is clear that the key to precision lies in the understanding of motivation, the ultimate reasons why people care about one thing but not another—why, say, they prefer a city with a park to a city alone. The goal is to join emotion with the rational analysis of emotion in order to create a deeper and more enduring conservation ethic.

Aldo Leopold, the pioneer ecologist and author of *A Sand County Almanac,* defined an ethic as a set of rules invented to meet circumstances so new or intricate, or else encompassing responses so far in the future, that the average person cannot foresee the final outcome. What is good for you and me at this moment might easily sour within ten years, and what seems ideal for the next few decades could ruin future generations. That is why any ethic worthy of the name has to encompass the distant future. The relationships of ecology and the human mind are too intricate to be understood entirely by unaided intuition, by common sense—that overrated capacity composed of the set of prejudices we acquire by the age of eighteen.

Values are time-dependent, making them all the more difficult to carve in stone. We want health, security, freedom, and pleasure for ourselves and our families. For distant generations we wish the same but not at any great personal cost. The difficulty created for the conservation ethic is that natural selection has programed people to think mostly in physiological time. Their minds travel back and forth across hours, days, or at most a hundred years. The forests may all be cut, radiation slowly rise, and the winters grow steadily colder, but if the effects are unlikely to become decisive for a few generations, very few people will be stirred to revolt. Ecological and evolutionary time, spanning centuries and millennia, can be conceived in an intellectual mode but has no immediate emotional impact. Only through an unusual amount of education and reflective thought do people come to respond emotionally to far-off events and hence place a high premium on posterity.

The deepening of the conservation ethic requires a greater measure of evolutionary realism, including a valuation of ourselves as opposed to other people. What do we really owe our remote descendants? At the risk of offending some readers I will suggest: Nothing. Obligations simply lose their meaning across centuries. But what do we owe ourselves in planning for them? Everything. If human existence has any verifiable meaning, it is that our passions and toil are enabling mechanisms to continue that existence unbroken, unsullied, and progressively secure. It is for ourselves, and not for them or any abstract morality, that we think into the distant future. The precise manner in which we take this measure, how we put it into words, is crucially important. For if the whole process of our life is directed toward preserving our species and personal genes, preparing for future generations is an expression of the highest morality of which human beings are capable. It follows that the destruction of the natural world in which the brain was assembled over millions of years is a risky step. And the worst gamble of all is to let species slip into extinction wholesale, for even if the natural environment is conceded more ground later, it can never be reconstituted in its original diversity. The first rule of the tinkerer, Aldo Leopold reminds us, is to keep all the pieces.

This proposition can be expressed another way. What event likely to happen during the next few years will our descendants most regret? Everyone agrees, defense ministers and environmentalists alike, that the worst thing possible is global nuclear war. If it occurs the entire human species is endangered; life as normal human beings wish to live it would come to an end. With that terrible truism acknowledged, it must be added that if no country pulls the trigger the worst thing that will *probably* happen — in fact is already well underway — is not energy depletion, economic collapse, conventional war, or even the expansion of totalitarian governments. As tragic as these catastrophes would be for us, they can be repaired within a few generations. The one process now going on that will take millions of years to correct is the loss of genetic and species diversity by the destruction of natural habitats. This is the folly our descendants are least likely to forgive us.

Extinction is accelerating and could reach ruinous proportions during the next twenty years. Not only are birds and mammals vanishing but such smaller forms as mosses, insects, and minnows. A conservative estimate of the current extinction rate is one thousand species a year, mostly from the destruction of forests and other key habitats in the tropics. By the 1990s the figure is expected to rise past ten thousand species a year (one species per hour). During the next thirty years fully one million species could be erased.

Whatever the exact figure—and the primitive state of evolutionary biology permits us only to set broad limits—the current rate is still the greatest in recent geological history. It is also much higher than the rate of production of new species by ongoing evolution, so that the net result is a steep decline in the world's standing diversity. Whole categories of organisms that emerged over the past ten million years, among them the familiar condors, rhinoceros, manatees, and gorillas, are close to the end. For most of their species, the last individuals to exist in the wild state could well be those living there today. It is a grave error to dismiss the hemorrhaging as a "Darwinian" process, in which species autonomously come and go and man is just the latest burden on the environment. Human destructiveness is something new under the sun. Perhaps it is matched by the giant meteorites thought to smash into the Earth and darken the atmosphere every hundred million years or so (the last one apparently arrived 65 million years ago and contributed to the extinction of the dinosaurs). But even that interval is ten thousand times longer than the entire history of civilization. In our own brief lifetime humanity will suffer an incomparable loss in aesthetic value, practical benefits from biological research, and worldwide biological stability. Deep mines of biological diversity will have been dug out and carelessly discarded in the course of environmental exploitation, without our even knowing fully what they contained.

The time is late for simple answers and divine guidance, and ideological confrontation has just about run its course. Little can be gained by throwing sand in the gears of industrialized society, even less by perpetuating the belief that we can solve any problem created by earlier spasms of human ingenuity. The need now is for a great deal more knowledge of the true biological dimensions of our problem, civility in the face of common need, and the style of leadership once characterized by Walter Bagehot as agitated moderation.

Ethical philosophy is a much more important subject than ordinarily conceded in societies dominated by religious and ideological orthodoxy. It faces an especially severe test in the complexities of the conservation problem. When the time scale is expanded to encompass ecological events, it becomes far more difficult to be certain about the wisdom of any particular decision. Everything is riddled with ambiguity; the middle way turns hard and general formulas fail with dispiriting consistency. Consider that a man who is a villain to his contemporaries can become a hero to his descendants. If a tyrant were to carefully preserve his nation's land and natural resources for his personal needs while keeping his people in poverty, he might unintentionally bequeath a rich, healthful environment to a reduced population for enjoyment in later, democratic generations. This caudillo will have improved the long-term welfare of his people by giving them greater resources and more freedom of action. The exact reverse can occur as well: today's hero can be tomorrow's destroyer. A popular political leader who unleashes the energies of his people and raises their standard of living might simultaneously promote a population explosion, overuse of resources, flight to the cities, and poverty for later generations. Of course these two extreme examples are caricatures and unlikely to occur just so, but they suffice to illustrate that, in ecological and evolutionary time, good does not automatically flow from good or evil from evil. To choose what is best for the near future is easy. To choose what is best for the distant future is also easy. But to choose what is best for both the near and distant futures is a hard task, often internally contradictory, and requiring ethical codes yet to be formulated.

AN ENDURING CODE of ethics is not created whole from absolute premises but inductively, in the manner of common law, with the aid of case histories, by feeling and consensus, through an expansion of knowledge and experience, influenced by the epigenetic rules of mental development, during which well-meaning and responsible people sift the opportunities and come to agree upon norms and directions.

The conservation ethic is evolving according to this pattern. It started centuries ago as a scattering of incidental thoughts and actions. The first biological preserves around the world were the by-products of selfish interests created, like most early art and learning, for the pleasure of the ruling classes. Among them were the gardens of the Kandy kings in Sri Lanka, the royal hunting reserves of Europe, and a few islands, such as Niihau in the Hawaiian group and Lignumvitae Key in Florida Bay, cordoned off for the use of private families.

I have visited all of these places, except Niihau, and many others as well, drawn by the opportunity offered for original biological research. In Cuba, on June 25, 1953, a month before Fidel Castro's assault on the Moncada barracks in Santiago de Cuba, I arrived on a far more modest mission in a jeep at a place called Blanco's Woods, near Cienfuegos. The tract was owned by a wealthy family who lived in Spain and declined to develop the land. All the surrounding forest had been cut down and converted into pasture and agricultural fields, leaving Blanco's Woods a rare refuge of native plants and animals of the coastal lowlands. To walk into that otherwise unprepossessing woodlot was to travel back into Cuba's geologic past, into the Pleistocene age before the coming of man—all thanks to what some would rightfully call the selfish actions of one family. Over 50 million years the Greater Antillean Islands, Cuba among them, had broken apart and drifted away from Central America eastward across the Caribbean Sea. In countless episodes the forests of Cuba were seeded with plants and animals from the mainland and surrounding islands. Many of the populations became extinct;

others hung on to evolve during thousands of generations into distinct genera and species, found nowhere else, woven together into intricate systems of competitors, predators, and prey. Biologists have given many of the organisms formal scientific names reflecting their origin and exclusive stronghold, such as *cubaensis, antillana, caribbaea,* and *insularis.* Now it has come down to this: in a negligible interval of evolutionary time, within the lifespan of Fidel Castro and one unheroic entomologist of approximately the same age visiting a nonstrategic part of the island, much of the woodland and hence a large part of Cuba's history have vanished. In 1953, on trial in Batista's court, Castro declared that history would absolve him. I wonder if it will, whether Blanco's Woods have since been cleared for the "good of the people"—meaning one or two generations—and to what degree the Cuban people will someday treasure such places as part of their national heritage, when heroes and political revolutions are dim in their memory.

Advances in conservation elsewhere in the world have been equally subordinate to whim and short-term social needs. The ginkgo tree, a relict of the ancient Asiatic forests and sole surviving species of an entire order of gymnospermous plants, was saved only because it was planted as an ornamental in Chinese and Japanese temple gardens over a period of centuries, long after it became extinct in the wild. Père David's deer held on for generations as an inhabitant of the imperial compound at Peking, after being hunted out over the rest of its once extensive range in China. In 1898, just before this final herd was destroyed, a new population was established by the Duke of Bedford on the grounds of Woburn Abbey. The stock has since been used to populate other reserves and parks. The great value of such by-the-fingernails species preservation is that it keeps alive the possibility of reconstituting original faunas and floras. Individuals can be transferred back to the original habitats and allowed to breed up to stable levels. Père David's deer itself may someday roam fresh in the relict woodlands of China.

Some kinds of organisms survive as the accidental beneficiaries of religion and magic. In Israel rare plants, largely exterminated in the surrounding agricultural land, grow in and around the Tel Dan, tombs of the holy men located near the sources of the Jordan River. When the biologist Michael J. D. White set out to analyze the genetic constitution of a group of interesting Australian grasshoppers called the Morabinae, he found them in sufficient numbers only in cemeteries and along railroad tracks. In the Western Ghats of India, sacred groves dating back to hunter-gatherer times today contain the best-preserved remnants of the original flora and fauna. Madhav Gadgil, one of India's foremost biologists and recipient of a gold medal in science from Prime Minister Indira Gandhi, has recommended that the groves serve as the nuclei of a system of national biotic reserves.

The modern practice of conservation has moved steadily forward from such primitive beginnings, but its philosophical foundations remain shaky. It still depends almost entirely on what may be termed surface ethics. That is, our relationship to the rest of life is judged on the basis of criteria that apply to other, more easily defined categories of moral behavior. This mode of reasoning is approximately the same as promoting literature because good writing helps to sell books, or art because it is useful for portraiture and scientific illustration. Of course the criteria are not in error—just spectacularly incomplete.

Thus we favor certain animals because they fill the superficial role of surrogate kin. It is the most disarming reason for nurturing other forms of life, and only a churl could find fault. Dogs are especially popular because they live by humanlike rituals of greeting and subservience. The family to whom they belong is part of their pack. They treat us like giant dogs, automatically alpha in rank, and clamor to be near us. We in turn respond warmly to their joyous greetings, tail wagging, slavering grins, drooped ears, groveling, bristling fur, and noisy indignation at territorial trespass. (Just as

I write this line I have to pause to calm down my own cocker spaniel, who is barking at a passing jogger. I say without thinking, "Quiet! Good *boy*!") The key to the compatibility of the two species is that dogs are descended with little behavioral modification from wolves. Like human beings, they and their wild cousins are happy carnivores, specialized to hunt large, swift, or otherwise unusually difficult prey in tightly coordinated groups. The wolf pack can catch mice and other small animals easily enough, but its real distinction is that it is also a superb instrument for bringing down a moose. The adaptation entails an extreme sensitivity to the moods of others. Dogs (domesticated wolves) are always ready for the communal hunt. They are primed to charge out the door with members of the human family in attendance, perhaps to chase down and slaughter a squirrel or rabbit, which, after an appropriate amount of fussing about and posturing in reconfirmation of status, will be shared with others. When not on the run or its equivalent (being carried along ecstatically in the family automobile), they follow the wolf's primal custom of spraying urine onto tree trunks and bushes (fireplugs and telephone poles will suffice) in order to mark out territory. At home, they metamorphose into children. The King Charles spaniel was bred to be an extreme specialist in this role. The adult possesses the small size, round head, and pug face of a puppy—also, let us acknowledge it frankly, of a baby—and is meant to be held in the lap.

Kinship affects emotion in other, unexpected ways. One of the most oddly disquieting events of my life was an encounter with Kanzi, a young pygmy chimpanzee. I was a guest of Sue Savage-Rumbaugh at the Language Research Center outside Atlanta, waiting in her office, when Kanzi was led in by a young woman who is helping to raise him. It was the first time I had seen this rare primate in life. I had a more than ordinary interest in it as an evolutionary biologist. The pygmy chimpanzee is arguably a distinct species from the ordinary chimpanzee. It appears to be somewhat less modified for arboreal existence than its sister species, and of the two it is the closer to man in certain key features of anatomy and behavior. The arms are longer and the legs shorter relative to the body. The head is more rounded, the forehead higher, and the jaw and brow less protruding. Overall the pygmy chimpanzee is remarkably similar in skeletal structure to "Lucy," the type specimen of *Australopithecus afarensis,* one of the probable direct precursors of man. It is the most humanlike of all animals. Its existence lends weight to the belief of many biologists that the evolutionary lines leading to human beings and chimpanzees split from a common stock in Africa as recently as five million years ago. There are also a few equally impressive similarities in behavior. The pygmy chimpanzee walks erect much of the time, and it learns many tasks more quickly and vocalizes more freely than the common chimpanzee. In sexual behavior it is closer than any other nonhuman primate to human beings. Females remain sexually receptive through most of their cycle, and they take a face-to-face position with the male in about a third of the couplings.

The pygmy chimpanzee is also endangered as a species. Wild populations are found only in one remote area in the Lomoko forest of Zaire, where a German lumber company has begun to conduct logging operations (in 1983, the time of writing). Only several dozen of the animals exist in captivity. Realizing the unique importance and threatened status of the species, scientists such as Savage-Rumbaugh, Adrienne Zihlman, and Jeremy Dahl are engaged in intensive studies of its biology and social behavior. Among the perhaps thirty million species of organisms on Earth, this is one that in my opinion deserves the highest priority in research and preservation.

Kanzi walked into the office and spotted me sitting in a chair on the far side of the room. He went into a frenzy of excitement, yelping and gesticulating to the two women with him in a way that seemed to exclaim, "That's a stranger! Why is he here? What are we going to do about him?" After a few minutes he calmed down and walked cautiously over to me, flicking glances from side to side as though plotting an emergency escape route. When he came near I brought my left hand up slowly and held it out, palm down and fingers slightly crumpled. I thought this was the very essence of humility and friendly intention, but he slapped my hand hard and backed off with a loud cry. The trainer murmured, "Oh, you're such a brave little boy!" (He *was* a brave little boy.) I didn't mind that my hand stung a bit. At that moment Kanzi's comfort and well-being seemed much more important than my own.

The trainer gave him a cup of grape juice, and he climbed into her lap to drink it and be cuddled. After a short wait he slid down to the floor and drifted back over to me. This time, having been coached by Sue Savage-Rumbaugh, I imitated the flutelike conciliatory call of the species, *wu-wu-wu-wu-wu* . . . with my lips pursed and what this time I believed to be a sincere, alert expression on my face. Now Kanzi reached out and touched my hand, nervously but gently, and stepped back a short distance to study me once again. The trainer gave me a cup of grape juice of my own. I flourished the cup as if offering a toast and took a sip, whereupon Kanzi climbed into my lap, took the cup, and drank most of the juice. Then we cuddled. Afterward everyone in the room had a good time playing ball and a game of chase with Kanzi.

The episode was unnerving. It wasn't the same as making friends with the neighbor's dog. I had to ask myself: was this really an animal? As Kanzi was led away (no farewells), I realized that I had responded to him almost exactly as I would to a two-year-old child—same initial anxieties, same urge to communicate and please, same gestures and food-sharing ritual. Even the conciliatory call was not very far off from the sounds adults make to comfort an infant. I was pleased that I had been accepted, that I had proved adequately human (was that the word?) and *sensitive* enough to get along with Kanzi.

WE ARE LITERALLY KIN to other organisms. The common and pygmy chimpanzees constitute the extreme case, the two species closest to human beings out of the contemporary millions. About 99 percent of our genes are identical to the corresponding set in chimpanzees, so that the remaining 1 percent accounts for all the differences between us. The chromosomes, the rodlike structures that carry the genes, are so close that only high-resolution photography and expert knowledge can tell many of them apart. Bishop Wilberforce's darkest thoughts might well be true; the creationists are justified in spending restless nights. The genetic evidence suggests that we resemble the chimpanzees in anatomy and a few key features of social behavior by virtue of a common ancestry. We descended from something that was more like a modern ape than a modern human being, at least in brain and behavior, and not very long ago by the yardstick of evolutionary time. Furthermore, the greater distances by which we stand apart from the gorilla, the orangutan, and the remaining species of living apes and monkeys (and beyond them other kinds of animals) are only a matter of degree, measured in small steps as a gradually enlarging magnitude of base-pair differences in DNA.

The phylogenetic continuity of life with humanity seems an adequate reason by itself to tolerate the continued existence of apes and other organisms. This does not diminish humanity—it raises the status of nonhuman creatures. We should at least hesitate before treating them as disposable matter. Peter Singer, a philosopher and animal liberationist, has gone so far as to propose that the circle of altruism be expanded beyond our own species to all animals with the capacity to feel and suffer, just as we have extended the label of brotherhood steadily until most people now feel comfortable with an all-inclusive phrase, the family of man. Christopher D. Stone, in *Should Trees Have Standing?,* has examined the legal implications of this enlarged generosity. He points out that until recently women, children, aliens, and members of minority groups had few or no legal rights in many societies. Although the policy was once accepted casually and thought congenial to the prevailing ethic, it now seems hopelessly barbaric. Stone asks why we should not extend similar protection to other species and to the environment as a whole. People still come first—humanism has not been abandoned—but the rights of owners should not be the exclusive yardstick of justice. If procedures and precedents existed to permit legal action to be taken on behalf of certain agreed-upon parts of the environment, the argument continues, humanity as a whole would benefit. I'm not sure I agree with this concept, but at the very least it deserves more serious debate than it has received. Human beings are a contractual species. Even religious dogma is hammered out as a system of mutual agreements. The working principles of ownership and privilege are arrived at by slow mutual consent, and legal theorists are a long way from having explored their limits.

If nobility is defined as reasoned generosity beyond expedience, animal liberation would be the ultimate ennobling act. Yet to force the argument entirely inside the flat framework of kinship and legal rights is to trivialize the case favoring conservation, to make it part of the surface ethic by justifying one criterion on the basis of another. It is also very risky. Human beings, for all their professed righteousness and brotherhood, easily discriminate against strangers and are content to kill them during wars declared for relatively frivolous causes. So it is much easier to find an excuse to exterminate another species. A stiffer dose of biological realism appears to be in order. We need to apply the first law of human altruism, ably put by Garrett Hardin: never ask people to do anything they consider contrary to their own best interests. The only way to make a conservation ethic work is to ground it in ultimately selfish reasoning—but the premises must be of a new and more potent kind.

An essential component of this formula is the principle that people will conserve land and species fiercely if they foresee a material gain for themselves, their kin, and their tribe. By this economic measure alone, the diversity of species is one of Earth's most important resources. It is also the least utilized. We have come to depend completely on less than 1 percent of living species for our existence, with the remainder waiting untested and fallow. In the course of history, according to estimates recently made by Norman Myers, people have utilized about 7,000 kinds of plants for food, with emphasis on wheat, rye, maize, and about a dozen other highly domesticated species. Yet at least 75,000 exist that are edible, and many of these are superior to the crop plants in use. The strongest of all arguments from surface ethics is a logical conclusion about this unrealized potential: the more the living world is explored and utilized, the greater will be the efficiency and reliability of the particular species chosen for economic use. Among the potential star species are these:

• The winged bean *(Psophocarpus tetragonolobus)* of New Guinea has been called a one-species supermarket. It contains more protein than cassava and potato and possesses an overall nutritional value equal to that of soybean. It is among the most rapidly growing of all plants, reaching a height of fifteen feet within a few weeks. The entire plant can be eaten, tubers, seeds, leaves, flowers, stems, and all, both raw and ground into flour. A coffeelike beverage can be made from the liquefied extract. The species has already been used to improve the diet in fifty tropical countries, and a special institute has been set up in Sri Lanka to study and promote it more thoroughly.

• The wax gourd *(Benincasa hispida)* of tropical Asia grows an inch every three hours over the course of four days, permitting multiple crops to be raised each year. The fruit attains a size of up to 1 by 6 feet and a weight of 80 pounds. Its crisp white flesh can be eaten at any stage, as a cooked vegetable, a base for soup, or a dessert when mixed with syrup.

• The Babussa palm *(Orbigyna martiana)* is a wild tree of the Amazon rain forest known locally as the "vegetable cow." The individual fruits, which resemble small coconuts, occur in bunches of up to 600 with a collective weight of 200 pounds. Some 70 percent of the kernel mass is composed of a colorless oil, used for margarine, shortening, fatty acids, toilet soap, and detergents. A stand of 500 trees on one hectare (2.5 acres) can produce 125 barrels of oil per year. After the oil has been extracted the remaining seedcake, which is about one-fourth protein, serves as excellent animal fodder.

Even with limited programs of research, biologists have compiled an impressive list of such candidate organisms in the technical literature. The vast majority of wild plants and animals are not known well enough (certainly many have not yet been discovered) even to guess at those with the greatest economic potential. Nor is it possible to imagine all the uses to which each species can be put. Consider the case of the natural food sweeteners. Several species of plants have been identified whose chemical products can replace conventional sugar with negligible calories and no known side effects. The katemfe *(Thaumatococcus danielli)* of the West African forests contains two proteins that are 1,600 times sweeter than sucrose and are now widely marketed in Great Britain and Japan. It is outstripped by the well-named serendipity berry *(Dioscoreophyllum cumminsii),* another West African native whose fruit produces a substance 3,000 times sweeter than sucrose.

Natural products have been called the sleeping giants of the pharmaceutical industry. One in every ten plant species contains compounds with some anticancer activity. Among the leading successes from the screening conducted so far is the rosy periwinkle, a native of the West Indies. It is the very paradigm of a previously minor species, with pretty five-petaled blossoms but otherwise rather ordinary in appearance, a roadside casual, the kind of inconspicuous flowering plant that might otherwise have been unknowingly consigned to extinction by the growth of sugarcane plantations and parking lots. But it also happens to produce two alkaloids, vincristine and vinblastine, that achieve 80 percent remission from Hodgkin's disease, a cancer of the lymphatic system, as well as 99 percent remission from acute lymphocytic leukemia. Annual sales of the two drugs reached $100 million in 1980.

A second wild species responsible for a medical breakthrough is the Indian serpentine root *(Rauwolfia serpentina)*. It produces reserpine, a principal source of tranquilizers used to relieve schizophrenia as well as hypertension, the generalized condition predisposing patients to stroke, heart malfunction, and kidney failure.

The natural products of plants and animals are a select group in a literal sense. They represent the defense mechanisms and growth regulators produced by evolution during uncounted generations, in which only organisms with the most potent chemicals survived to the present time. Placebos and cheap substitutes were eliminated at an early stage. Nature has done much of our work for us, making it far more efficient for the medical researcher to experiment with extracts of living tissue than to pull chemicals at random off the laboratory shelf. Very few pharmaceuticals have been invented from a knowledge of the first principles of chemistry and medicine. Most have their origin in the study of wild species and were discovered by the rapid screening of large numbers of natural products.

For the same reason, technical advances utilizing natural products have been achieved in many categories of industry and agriculture. Among the most important have been the development of phytoleum, new plant fuels to replace petroleum; waxes and oils produced from indefinitely renewing sources at more economical rates than previously thought possible; novel kinds of fibers for paper manufacture; fast-growing siliceous plants, such as bamboo and elephant grass, for economical dwellings; superior methods of nitrogen fixation and soil reclamation; and magic-bullet techniques of pest control, by which microorganisms and parasites are set loose to find and attack target species without danger to the remainder of the ecosystem. Even the most conservative extrapolation indicates that many more such discoveries will result from only a modest continuing research effort.

Furthermore, the direct harvesting of free-living species is only a beginning. The favored organisms can be bred over ten to a hundred generations to increase the quality and yield of their desired product. It is possible to create strains that do well in new climates and the special environments required for mass production. The genetic material comprising them is an additional future resource; it can be taken apart gene by gene and distributed to other species. Thomas Eisner, one of the pioneers of chemical ecology, has used a striking analogy to explain these two levels of utilization of wild organisms. Each of the millions of species can be visualized as a book in a library. No matter where it originates, it can be transferred and put to use elsewhere. No matter how rare in its original state, it can be copied many times over and disseminated to become indefinitely abundant. An orchid down to the last hundred individuals in a remote valley of the Peruvian Andes, which also happens to be the source of a medicinal alkaloid, can be saved, cultured, and converted into an important crop in gardens and greenhouses around the world. But there is much more to the species than the alkaloid or other useful material that it happens to package. It is not really a conventional book but more like a looseleaf notebook, in which the genes are the equivalent of detachable pages. With new techniques of genetic engineering, biologists will soon be able to lift out desirable genes from one species or strain and transfer them to another. A valuable food plant, for example, can be given DNA from wild species conferring biochemical resistance to its most destructive disease. It can be altered by parallel procedures to grow in desert soil or through longer seasons.

A notable case in point is the primitive form of maize, *Zea diploperennis,* recently discovered in a mountain forest of southwestern Mexico. It is still known from three patches covering a mere ten acres (at any time a bulldozer might easily have extinguished the entire species, within hours). *Zea diploperennis* possesses genes for perennial growth, making it unique among all other known varieties of corn. It is thus the potential source of a hereditary trait that could reduce growing time and labor costs, making cultivation more feasible in ecologically marginal areas.

There are few countries in the world that do not harbor unique species and genetic strains still unknown to the people who live there. There is no country that would fail to benefit from the importation of such undiscovered organisms. With these facts in mind I find it astonishing that so little attention is being given to the exploration of the living world. The set of disciplines collectively called evolutionary biology, including initial field surveys, taxonomy, ecology, biogeography, and comparative biochemistry, remains among the most poorly funded in science. The amount spent globally in 1980 on such research in the tropics, where the great majority of organisms live, was $30 million—somewhat less than the cost of two F-15 Eagle fighter-bombers, approximately 1 percent of the grants for health-related research in the United States, or a few weeks' liquor bill for the populace of New York City.

Let us postpone for the moment moral arguments of the conventional kind. It would be to the direct economic advantage of most governments to invest more in the study of their own living resources. Because evolutionary biology exists so close to the poverty line, it offers society what economists call increasing returns to scale: a modest absolute expenditure in dollars will yield large relative benefits. The reason is that the existing low level of activity causes most opportunities to remain unmet, with the result that the marketplace stays largely empty. Museums, meant at their founding to be national research centers, are everywhere understaffed. Taxonomy, the principal occupation of museum scientists, is a declining profession through lack of support. The neglect is all the more puzzling because the value of the research is widely appreciated within the scientific community. Any biologist who tries to get an identification of an organism in order to facilitate its further study knows that he may be in for a long wait. Even when the research has considerable economic potential, it is often at risk because of delays and inadequate data.

The diversity of species is so immense that the Linnaean enterprise of describing the living world remains by force a part of modern science. In addition to more and better staffed museums, we (scientists, individual countries, the world) would benefit from institutes for the extended study of the organisms once they have been classified. There the previously unknown species can be screened for economic and medical potential, their ecology and physiological traits probed. The accumulating data will also reveal the complex processes by which species originate and go extinct, information needed to guide the practice of conservation.

A few such institutions of high quality exist today, among them Brazil's National Institute for Research on Amazonia, the Marine Biological Laboratory at Woods Hole, Massachusetts, and the Smithsonian Tropical Research Institute in Panama. But even if these pioneering organizations were operated at full current capacity, they could handle only a minute fraction of the different kinds of organisms around the world. The most urgent need is for an increased research capacity in the tropics, where perhaps 90 percent of species exist.

I will now add a note of optimism that I know is shared by many biologists. The exploration of natural resources is the kind of research most readily justified in the underdeveloped countries, especially those in the tropics. It is also the kind they can most easily afford. These nations occasionally need accelerators, satellites, mass spectrometers, and the other accouterments of big science, but such equipment can be borrowed during cooperative ventures with the richer countries. The economically less developed countries can do better with skilled and semiskilled workers who make expeditions into the wild, collect and prepare specimens, culture promising varieties, and spend the long hours of close observation needed to understand growth and behavior. This kind of science is labor-intensive, best performed by people who love the land and organisms for their own sake. Its results will gain worldwide recognition and serve as a source of national pride.

Can there be an Ecuadoran biology, a Kenyan biology? Yes, if they focus on the uniqueness of indigenous life. Will such efforts be important to international science? Yes, because evolutionary biology is a discipline of special cases woven into global patterns. Nothing makes sense except in the light of the histories of local faunas and floras. It is further true that all of biology, from biochemistry to ecology, is moving toward a greater emphasis on evolution and its resultant particularity.

Finally, the efforts of generations to come will be frustrated unless they are safeguarded with national reserve systems of the kind recently pioneered by Brazil, Costa Rica, and Sri Lanka, where the parcels of land set aside are chosen to achieve a maximum protection of organic diversity. Otherwise hundreds of species will continue to vanish each year without so much as the standard double Linnaean names to record their existence. Each takes with it millions of bits of genetic information, a history ages long, and potential benefits to humanity left forever unmeasured.

TO SUMMARIZE: a healthful environment, the warmth of kinship, right-sounding moral strictures, sure-bet economic gain, and a stirring of nostalgia and sentiment are the chief components of the surface ethic. Together they are enough to make a compelling case to most people most of the time for the preservation of organic diversity. But this is not nearly enough: every pause, every species allowed to go extinct, is a slide down the ratchet, an irreversible loss for all. It is time to invent moral reasoning of a new and more powerful kind, to look to the very roots of motivation and understand why, in what circumstances and on which occasions, we cherish and protect life. The elements from which a deep conservation ethic might be constructed include the impulses and biased forms of learning loosely classified as biophilia. Ranging from awe of the serpent to the idealization of the savanna and the hunter's mystique, and undoubtedly including others yet to be explored, they are the poles toward which the developing mind most comfortably moves. And as the mind moves, picking its way through the vast number of choices made during a lifetime, it grows into a form true to its long, unique evolutionary history.

I have argued in this book that we are human in good part because of the particular way we affiliate with other organisms. They are the matrix in which the human mind originated and is permanently rooted, and they offer the challenge and freedom innately sought. To the extent that each person can feel like a naturalist, the old excitement of the untrammeled world will be regained. I offer this as a formula of reenchantment to invigorate poetry and myth: mysterious and little known organisms live within walking distance of where you sit. Splendor awaits in minute proportions.

Why then is there resistance to the conservation ethic? The familiar argument is that people come first. After their problems have been solved, we can enjoy the natural environment as a luxury. If that is indeed the answer, the wrong question was asked. The question of importance concerns purpose. Solving practical problems is the means, not the purpose. Let us assume that human genius has the power to thread the needles of technology and politics. Let us imagine that we can avert nuclear war, feed a stabilized population, and generate a permanent supply of energy—what then? The answer is the same all around the world: individuals will strive toward personal fulfillment and at last realize their potential. But what is fulfillment, and for what purpose did human potential evolve?

The truth is that we never conquered the world, never understood it; we only think we have control. We do not even know why we respond a certain way to other organisms, and need them in diverse ways, so deeply. The prevailing myths concerning our predatory actions toward each other and the environment are obsolete, unreliable, and destructive. The more the mind is fathomed in its own right, as an organ of survival, the greater will be the reverence for life for purely rational reasons.

Natural philosophy has brought into clear relief the following paradox of human existence. The drive toward perpetual expansion—or personal freedom—is basic to the human spirit. But to sustain it we need the most delicate, knowing stewardship of the living world that can be devised. Expansion and stewardship may appear at first to be conflicting goals, but they are not. The depth of the conservation ethic will be measured by the extent to which each of the two approaches to nature is used to reshape and reinforce the other. The paradox can be resolved by changing its premises into forms more suited to ultimate survival, by which I mean protection of the human spirit.

Surinam

ETERNAL SURINAM: the image of the land I kept for many years symbolized the tangle of dreams and boyhood adventures from which I had originally departed, the home country of all naturalists, and the quiet refuge from which personal beliefs might someday be redeemed in a permanent and more nearly perfect form. It is appropriate, then, to describe the reality of that particular place before returning a final time to its image.

Surinam is a sovereign country with a fertile coastal plain, interior wilderness, and one of the richest forest reserves in the world. It is often called the ornithologist's paradise for the variety of neotropical bird species seen more easily there than in most of the rest of South America. Parrots flock among the palms within the city limits of Paramaribo. Over a hundred kinds of hummingbirds and cotingas flash through the flowering canopies of the nearby forests. A short drive and boat trip to the south will bring you to guans, tinamous, manakins, bellbirds, ant-thrushes, and toucans, and perhaps provide a glimpse of the harpy eagle, the giant predator of monkeys and sloths and apex of the arboreal energy pyramid. It is a general rule that, when the bird fauna stays intact, so does the rest of the fauna and flora. The interior of Surinam is a fragment of tropical America as it was ten thousand years ago, or at least approximately so, when the first Indian colonists walked in from the Panamanian land bridge.

Location: north coast of South America, bracketed by French Guiana to the east and Guyana to the west, with Brazil sharing the southern border. Population: 350,000, mostly concentrated on the coast, especially in and around Paramaribo. Agriculture is mixed and moderately successful, with emphasis on rice as the principal export crop. One of the largest hydroelectric plants in South America is located at the Brokopondo Dam; it delivers the bulk of the power used by a highly productive bauxite operation, still mostly foreign-owned. The Surinamese people are courteous and friendly, adding considerably to the potential of tourism as an economic resource. They react with special warmth to visitors who struggle with Takki-Takki, the national Creole dialect, although Dutch or English will serve you well almost anywhere in the country.

Climate: sweltering. Education: valued and improving. Roads: few. The Netherlands gave Surinam independence in 1975 and a promised allowance of $100 million annually for fifteen years. By 1982 the per capita income was $2,500, one of the highest among developing countries. One in three persons owned an automobile, while refrigerators and television sets were routinely stocked in private homes. The long-term future seems bright for this little country, which has a bountiful environment, a small population, and hence a period of grace that was not granted to most of the Third World when colonialism came to an end.

Bernhardsdorp has changed strikingly since my visit in 1961. Touched at last by the population sprawl originating in Paramaribo and Lelydorp, it grew from a tiny Arawak village into a town of about five hundred people of Javanese, Chinese, Amerindian, and Creole ancestry, an ethnic microcosm of larger Surinam. Today the scene is classic tropical-rural. The thatched huts are outnumbered by conventional one- or two-room dwellings built on basement pilings from plank sidings and sheetmetal roofs. The lush pastureland and gardens, crisscrossed by drainage ditches, yield an abundance of vegetables, dairy products, and poultry for local consumption and nearby markets. In the center of town, by the main dirt road, is a small store run by a Chinese family. Someone has erected Coca Cola signs and a billboard with the national coat of arms, featuring two armed Arawak warriors, a circular shield emblazoned with sailing ship, star, and palm tree, and beneath these figures, on the flying scroll, the motto "Justitia, Pietas, Fides." The bulldozer came: the forest has been mostly cleared, leaving behind a scattering of palms and second-growth edge thickets. There is also a tall tree with tear-shaped oropendola nests hanging in military rows beneath its horizontal branches. The town is not yet on any map I have been able to find. A carefully lettered sign at the turnoff from the paved Lelydorp-Zanderij road proudly proclaims its existence: BERNHARDSDORP.

In 1980 all this bright picture was darkened by the advent of barbarism. The democratically elected government of Henk Arron was overthrown by Revolutionary Leader Dési Bouterse, a military physical-education instructor with scant education. At first suspicious of socialism, but then schooled in Marxism-Leninism by his teacher and mistress, Bouterse drifted leftward and began to court Fidel Castro and the Soviet Union. In December 1982 Bouterse, without warning, ordered the arrest and execution of fifteen of the country's leading citizens, including lawyers, journalists, and union leaders. The next morning all but one were dead. With a substantial fraction of the old leadership erased and hundreds of citizens soon joining the tens of thousands already in exile, Bouterse announced "the building of a new Surinam."

As I write, it is a Surinam of silence and fear. The modest tourist trade has ended, aid from the Netherlands and United States has been cut off, unemployment is rising, and the once substantial foreign reserves are quickly drying up. The state university has been closed, and the key radio stations and union headquarters have been either burned or blown up. Plainclothes police arrest citizens randomly for questioning. People rarely speak about the government for fear of the informers they now believe to be everywhere. In the words of one exile, Surinam has become a "country of mutes." Its anxiety is the projected mood of a frightened, paranoid ruler. Rumors are circulating of a planned military coup to overthrow Bouterse, perhaps backed clandestinely by the United States. This has of course been denied. Cuba has been rebuffed, its ambassador expelled, through fear of a coup from the left. In contrast, Brazil has managed to expand contact with Bouterse's regime with the expressed hope of meliorating it. All are wrestling with a problem as old as recorded history: how to deal with the kingdoms of Caliban.

There is a way for the mind to find some ease in such matters. Whatever its denouement the Bouterse episode, the national tragedy, can be seen as no more than a tick in what will be the ultimate history of Surinam. Its people will survive to see ecological and then evolutionary change, within which biography and political events become cyclical and shrink steadily in proportion.

The swiftness of human change and the transience of power are well remembered in the words of a man who saw it all on a grander stage and far enough back in time to gain convincing authority: the wise Stoic emperor Marcus Aurelius. Take the distant view, he said, and observe that those who praise and those praised both endure for a short season and are gone, and "all this too in a tiny corner of this continent, and not even there are all in accord, no nor a man with himself."

> The persons men wish to please, the objects they wish to gain, the means they employ—think of the character of all these! How soon will Time hide all things! How many a thing has it already hidden!

I wish I could ask him: Marcus Aurelius, do you agree that tragedy, like value, is dependent upon the scale of time? If you could be a philosopher king in this century, and sail to some new Ionia in search of wisdom, would you turn to conservation? Is it possible that humanity will love life enough to save it?

I will remember Bernhardsdorp as a special place, a portal to far-reaching dreams. To the south stretches Surinam eternal, Surinam serene, a living treasure awaiting assay. I hope that it will be kept intact, that at least enough of its million-year history will be saved for the reading. By today's ethic its value may seem limited, well beneath the pressing concerns of daily life. But I suggest that as biological knowledge grows the ethic will shift fundamentally so that everywhere, for reasons that have to do with the very fiber of the brain, the fauna and flora of a country will be thought part of the national heritage as important as its art, its language, and that astonishing blend of achievement and farce that has always defined our species.

Reading Notes

Epigraph (p. i)

"Soft, to Your Places," by Thomas Kinsella, in *Selected Poems, 1956–1968* (Dublin: Dolmen Press, 1973).

Prologue

I first used the term *biophilia* in an article by that title in the *New York Times Book Review,* January 14, 1979, p. 43.

Bernhardsdorp

Leo Marx, *The Machine in the Garden: Technology and the Pastoral Ideal in America* (New York: Oxford University Press, 1964).

Yi-Fu Tuan, *Topophilia: A Study of Environmental Perception, Attitudes, and Values* (Englewood Cliffs: Prentice-Hall, 1974).

The remarkable attack mechanism of the oomycete fungus *Haptoglossa mirabilis* was recently worked out in detail by E. Jane Robb and G. L. Barron, "Nature's Ballistic Missile," *Science,* 218:1221–1222 (1982).

The density of soil organisms is based on estimates given in John A. Wallwork, *The Distribution and Diversity of Soil Fauna* (New York: Academic Press, 1976), and Peter H. Raven, Ray F. Evert, and Helena Curtis, *Biology of Plants,* 3rd ed. (New York: Worth Publishers, 1981).

The estimates of the number of nucleotide pairs in various kinds of organisms are given in the authoritative review by Ralph Hinegardner, "Evolution of Genome Size," pp. 179–199 in *Molecular Evolution,* ed. F. J. Ayala (Sunderland, Mass.: Sinauer Associates, 1976). Four permutations exist at the nucleotide pair sites, AT, TA, CG, and GC; and the amount of information present at each site can be roughly approximated as $\log_2 4 = 2$ bits. The number of bits per English word was derived by Henry Quastler in "A Primer on Information Theory," pp. 3–49 in *Symposium on Information Theory in Biology,* ed. H. P. Yockey (New York: Pergamon Press, 1958).

The upper estimate of 30 million living species of insects may seem improbable at first, but it has been carefully argued and documented by Terry L. Erwin, "Tropical Forest Canopies: The Last Biotic Frontier," *Bulletin of the Entomological Society of America,* 29:14–19 (1983).

The Superorganism

An authoritative account of the Critical Size Project is given by Sam Iker, "Islands of Life in a Forest Sea," *Mosaic* (National Science Foundation, Washington), 13:25–30 (September–October 1982). A nicely illustrated but very brief account is also presented by Peter T. White, "Tropical Rain Forests: Nature's Dwindling Treasures," *National Geographic,* 163:2–47 (January 1983).

I have never witnessed the spontaneous crash of a forest tree, but I have seen many giant rain forest trees brought down by the chain saw. That experience was used to reconstruct what must occur in the Amazonian forest.

The general biology of leafcutter ants is described by Edward O. Wilson, *The Insect Societies* (Cambridge: Harvard University Press, 1971), and Neal A. Weber, *Garden Ants: The Attines* (Philadelphia: American Philosophical Society, 1972).

The Time Machine

The account of the conversation between Louis Agassiz and Benjamin Peirce is in A. Hunter Dupree's *Asa Gray* (Cambridge: Harvard University Press, 1959). We know the subject but of course not the exact words. On the other hand, Agassiz's remark about Darwinism that evening ("We must stop this") is just as Gray later recalled it.

The exchanges between Agassiz, Darwin, and their friends are taken from David L. Hull's admirable *Darwin and His Critics: The Reception of Darwin's Theory of Evolution by the Scientific Community* (Cambridge: Harvard University Press, 1973). The protest by Agassiz is quoted from an article published posthumously in the *Atlantic Monthly*, 1874.

Bertrand Russell is quoted from an interview reprinted in *The Humanist*, November–December 1982, p. 39.

The characterization of restrictionists and expansionists is elaborated by Loren R. Graham in *Between Science and Values* (New York: Columbia University Press, 1981).

Great effects do not imply great causes: I first heard the key conclusion of Darwin expressed in just this way by the philosopher John Passmore at a lecture at Cambridge University in 1982.

The Darwin notebooks are cited by P. H. Barrett, *Metaphysics, Materialism, and the Evolution of Mind: Early Writings of Charles Darwin* (Chicago: University of Chicago Press, 1980).

For examples of the modern restrictionist view of science, see John W. Bowker, "The Aeolian Harp: Sociobiology and Human Judgment," *Zygon*, 15:307–333(1980); Theodore Roszak, "The Monster and the Titan: Science, Knowledge, and Gnosis," *Daedalus*, 103:17–32 (1974); and William Irwin Thompson, *The Time Falling Bodies Take to Light* (New York: St. Martin's Press, 1981).

The Bird of Paradise

The description of the Emperor of Germany bird of paradise is based on my examination of specimens in the Museum of Comparative Zoology, Harvard University, and the excellent painting and biological summary by William T. Cooper and Joseph M. Forshaw in *The Birds of Paradise and Bower Birds* (Boston: David R. Godine, 1977). I never saw the species in the wild, even though in 1955 I walked over substantial parts of the Huon Peninsula inland from

Finschhafen and Lae, and many *Paradisaea guilielmi* probably saw me. The reason is very simple: I was studying ants, encountering over 300 different kinds in this area alone, and almost always had my gaze focused on the ground. On one occasion I heard a sharp cry high in the treetops, and an Australian biologist nearby shouted "Bird of paradise!" But by the time I could adjust my eyeglasses and look up, the bird was gone.

The Poetic Species

David Hilbert wrote on the vital importance of perpetual discovery in the preamble to his celebrated article presenting twenty-three fundamental problems in modern mathematics, "Sur les problèmes futurs des mathématiques," in *Compte rendu du Deuxième Congrés International des Mathématiciens* (Paris, 1900), pp. 58–114.

Einstein on Planck: "Principles of Research," *Ideas and Opinions by Albert Einstein,* based on *Mein Weltbild,* ed. Carl Seelig; rev. Sonja Bargmann (New York: Bonanza Books, 1954).

Dirac wrote on the relationship between beauty and scientific truth in "The Evolution of the Physicist's Picture of Nature," *Scientific American,* 208:45–53 (May 1963). Weyl on aesthetics and truth: as quoted from a conversation with Freeman J. Dyson in an obituary essay, *Nature,* 177:457–458 (1956).

Hilbert's remarks are quoted by William N. Lipscomb in "Aesthetic Aspects of Science," pp. 1–24 in *The Aesthetic Dimension of Science,* ed. Deane W. Curtin (New York: Philosophical Library, 1982).

I am indebted to the following sources for a more formal concept of art and the humanities used in my comparison with science: Richard W. Lyman et al., *The Humanities in American Life,* Report of the Commission on the Humanities (Berkeley: University of California Press, 1980); W. Jackson Bate, "The Crisis in English Studies," *Harvard Magazine,* September–October 1982, pp. 46–53; and Paul Oskar Kristeller, "The Humanities and Humanism," *Humanities Report,* January 1982, pp. 17–18.

Roger Shattuck on the autonomous tradition of art: "Humanizing the Humanities," *Change*, November 1974, pp. 4–5.

T. S. Eliot wrote on the discipline of the poet in "Tradition and the Individual Talent" (1919), in *Selected Prose of T. S. Eliot* (New York: Harcourt Brace Jovanovich, 1975).

Octavio Paz's "The Broken Waterjar" is translated by Lysander Kemp in *Early Poems, 1935–1955.*

Some of the best testimony concerning the creative process is to be found in lectures given by scientists and other scholars in the annual Nobel Conferences arranged by the faculty of Gustavus Adolphus College. The most pertinent are: *Creativity*, ed. John D. Roslansky (Amsterdam: North Holland, 1970); *The Aesthetic Dimension of Science*, ed. Deane W. Curtin (New York: Philosophical Library, 1982); and *Mind in Nature*, ed. Richard Q. Elvee (New York: Harper and Row, 1982).

Cyril S. Smith recounts the origins of his love for metallurgy in *A Search for Structure: Selected Essays on Science, Art, and History* (Cambridge: MIT Press, 1981).

Camus characterizes the creative detour to rediscover the images of childhood in the preface to *The Wrong Side and the Right Side*, reprinted in *Lyrical and Critical Essays* (New York: Alfred A. Knopf, 1969).

Hideki Yukawa presented his view of the central role of analogy in *Creativity and Intuition: A Physicist Looks East and West*, trans. John Bester (Tokyo: Kodansha International, 1973). Einstein on analogies: "It is easy to find a superficial analogy which really expresses nothing. But to discover some essential feature hidden beneath the surface of external differences [and] to form on this basis a new successful theory is a typical example of the achievement of a successful theory by means of a deep and fortunate analogy." *The Evolution of Physics* (New York: Simon and Schuster, 1938).

Robert H. MacArthur and I published our principal work in "An Equilibrium Theory of Insular Biogeography," *Evolution*,

17:373–387 (1963), and more fully in *The Theory of Island Biogeography* (Princeton: Princeton University Press, 1967). A more recent and comprehensive account of the theory and related topics is given by Mark Williamson in *Island Populations* (Oxford: Oxford University Press, 1981).

Bishop Lowth is quoted and the importance of his analysis examined by M. H. Abrams in *The Mirror and the Lamp* (New York: Oxford University Press, 1953), an authoritative review of the romantic tradition and the origins of literary criticism.

Richard Rorty describes humanity as the poetic species in his superb review of the philosophy of mind: "For beyond the vocabularies useful for prediction and control—the vocabulary of natural science—there are the vocabularies of our moral and our political life and of the arts, of all those human activities which are not aimed at prediction and control but rather in giving us self-images which are worthy of our species. Such images are not true to the nature of species or false to it, for what is really distinctive about us is that we can rise above questions of truth or falsity. We are the poetic species, the one which can change itself by changing its behavior—and especially its linguistic behavior, the words it uses." "Mind as Ineffable," pp. 60–95 in *Mind in Nature,* ed. R. Q. Elvee (New York: Harper and Row, 1982).

An excellent account of cave art and its possible use in the transmission of culture has been provided by John E. Pfeiffer, *The Creative Explosion: An Inquiry into the Origins of Art and Religion* (New York: Harper and Row, 1982).

Kinsella's "Midsummer," *Selected Poems, 1956–1968* (Dublin: Dolmen Press, 1973).

Eberhart's stanza is from "Ultimate Song," *Collected Poems, 1930–1976* (New York: Oxford University Press, 1976).

Some of the key reference works and textbooks on the mind and memory, including the node-link model, are *Cognitive Psychology and Its Implications,* by John R. Anderson (San Francisco: W. H. Freeman, 1980); *Mechanics of the Mind,* by Colin Blakemore (New York: Cambridge University Press, 1977); *Brainstorms: Philosophical Essays on Mind and Psychology,* by Daniel C. Dennett (Montgomery, Vt.: Bradford Books, 1978); *Psychology,* by Gardner Lindzey, C. S. Hall, and R. F. Thompson (New York: Worth Publishers, 1975); *Human Memory: The Processing of Information,* by G. R. and Elizabeth F. Loftus (Hillsdale: Lawrence Erlbaum Associates, 1976); *The Psychobiology of Mind,* by William R. Uttal (Hillsdale: Lawrence Erlbaum Associates, 1978); and *Cognitive Psychology,* by Wayne A. Wickelgren (Englewood Cliffs: Prentice-Hall, 1979).

The measurement of varying brain arousal by different geometric designs was reported by Gerda Smets in *Aesthetic Judgment and Arousal: An Experimental Contribution to Psycho-physics* (Leuven, Belgium: Leuven University Press, 1973).

Stella is quoted and his work analyzed in J. Gray Sweeney's *Themes in American Painting* (published under the auspices of the Grand Rapids Art Museum, Michigan, 1977).

The Serpent

I have drawn most of the facts on the serpent in culture from Balaji Mundkur's *The Cult of the Serpent: An Interdisciplinary Survey of Its Manifestations and Origins* (Albany: State University of New York Press, 1983). This is a highly original and masterly work. Although I have long thought about our awe of the serpent, Mundkur has documented it in impressive detail from the history of art and literature.

A detailed and authoritative account of Zeus Meilikhios and the snake-Erinyes is given by Jane Ellen Harrison, *Prolegomena to the Study of Greek Religion,* 3rd ed. (Cambridge: Cambridge University Press, 1922).

The conception of biasing in mental development and its relation to human nature and culture is presented in greater detail in Charles J. Lumsden and Edward O. Wilson, *Promethean Fire* (Cambridge: Harvard University Press, 1983).

The Right Place

José Ortega y Gasset, *Meditations on Hunting,* trans. Howard B. Wescott (New York: Charles Scribner's Sons, 1972). Other excellent discussions of the hunter's mystique are given by Paul Shepard, *The Tender Carnivore and the Sacred Game* (New York: Charles Scribner's Sons, 1973), and John G. Mitchell, *The Hunt* (New York: Alfred A. Knopf, 1980).

The pygmy desmognath salamander I collected was *Desmognathus chermocki.* It has since been formally combined with the more widespread *Desmognathus aeneus,* although I am informed by one of its discoverers, Barry D. Valentine, that its status remains problematical. In either case the field observations made in Alabama retain their significance with respect to the behavioral diversity of the desmognaths.

William Mann's account of ant collecting in Cuba is in "Stalking Ants Savage and Civilized," *National Geographic,* 66:171–192 (August 1934).

The basic research on orientation and habitat selection in bacteria is ably summarized by Daniel E. Koshland, Jr., *Bacterial Chemotaxis as a Model Behavioral System* (New York: Raven Press, 1980).

The evidence for the savanna habitat as the home of early man has been presented by several authors, including Karl W. Butzer, "Environment, Culture, and Human Evolution," *American Scientist,* 65:572–584 (1977), and Glynn Isaac, "Casting the Net Wide: A Review of Archaeological Evidence for Early Hominid Land-Use and Ecological Relations," pp. 114–134 in *Current Arguments on Early Man,* ed. L.-K. Königsson (New York: Pergamon Press, 1980).

Gordon H. Orians developed the idea of the psychologically optimum human environment in "Habitat Selection: General Theory and Applications to Human Behavior," pp. 49–66 in *The Evolution of Human Social Behavior,* ed. Joan S. Lockard (New York: Elsevier North Holland, 1980). The diary entries by Marcy and Parker are in Public Document 577 of the 31st Congress (1849); quoted by Orians.

The metaphor of the cataract of sand is given in the first chapter of *Moby Dick.* Herman Melville understood as few other authors the innate aesthetic sense of the environment and especially the compelling attraction of open water: "Say, you are in the country, in some high land of lakes. Take almost any path you please, and ten to one it carries you down in a dale, and leaves you there by a pool in the stream. There is magic in it. Let the most absent-minded of men be plunged in his deepest reveries — stand that man on his legs, set his feet a-going, and he will infallibly lead you to water, if water there be in all that region." The yearning is of a very general kind, generating symbolism across many categories of thought. "It is the image of the ungraspable phantom of life; and this is the key to it all."

Cyril S. Smith, *A Search for Structure: Selected Essays on Science, Art, and History* (Cambridge: MIT Press, 1981), p. 355.

On the colonization of space: the concept of self-contained stations was first brought into public discussion by Gerard K. O'Neill in an article for *Physics Today* (1974) and developed at length in his book *The High Frontiers: Human Colonies in Space* (New York: Bantam Books, 1976). An excellent popular exposition is also provided by T. A. Heppenheimer in *Colonies in Space* (Harrisburg: Stackpole Books, 1977). Extensions and criticisms, some of the latter quite severe, were written by physicists, ecologists, and others for *Space Colonies,* ed. Stewart Brand (New York: Penguin Books, 1977).

The Conservation Ethic

Aldo Leopold, "The Land Ethic," *A Sand County Almanac and Sketches Here and There* (New York: Oxford University Press, 1949).

The acceleration of species extinction and its dangers for mankind have been ably documented by Norman Myers, *The Sinking Ark* (Elmsford: Pergamon Press, 1979), and Paul R. and Anne Ehrlich, *Extinction: The Causes and Consequences of the Disappearance of Species* (New York: Random House, 1981). They have been examined still further by Peter H. Raven and others in three National Research Council reports: *Conversion of Tropical Moist Forests* (1980); *Research Priorities in Tropical Biology* (1980); and *Ecological Aspects of Development in the Humid Tropics* (1982).

For information on the rare flora of Tel Dan (Tel el Kadi), in the Hule Valley of Israel, I am grateful to Jehoshua Kugler and Eviatar

Nevo. The role of the sacred groves as unplanned nature reserves is explained by Madhav Gadgil and V. D. Vartak, "The Sacred Groves of Western Ghats in India," *Economic Botany,* 30:152–160 (1974).

Perhaps the best historical review of the origin of the conservation ethic in the United States is Donald Fleming's "Roots of the New Conservation Movement," pp. 7–91 in *Perspectives in American History,* vol. 6, ed. Donald Fleming and Bernard Bailyn (Lunenburg: Stinehour Press, for the Charles Warren Center for Studies in American History, Harvard University, 1972). The concept of wilderness in particular is explored by Roderick Nash in his classic *Wilderness and the American Mind,* rev. ed. (New Haven: Yale University Press, 1973).

The idea of extended kinship contributing to the conservation ethic has been systematically examined by Gordon M. Burghardt and Harold A. Herzog, Jr., "Beyond Conspecifics: Is Brer Rabbit Our Brother?," *BioScience,* 30:763–768 (1980).

The biology and status of the pygmy chimpanzee is described in "An Uncommon Chimp," by Paul Raeburn, *Science 83,* 4:40–48 (June 1983).

Peter Singer, *The Expanding Circle: Ethics and Sociobiology* (New York: Farrar, Straus and Giroux, 1981). Christopher D. Stone, *Should Trees Have Standing? Toward Legal Rights for Natural Objects* (Los Altos: William Kaufmann, 1974).

Garrett Hardin's tough-love approach to ethical philosophy is concisely expressed in *The Limits of Altruism: An Ecologist's View of Survival* (Bloomington: Indiana University Press, 1977).

My examples of edible tropical plants are taken from Norman Myers' important encyclopedic account, *A Wealth of Wild Species: Storehouse for Human Welfare* (Boulder: Westview Press, 1983).

Thomas Eisner compared species to a genetic looseleaf notebook in his testimony on the Endangered Species Act; his prepared statement was published in *The Congressional Record,* vol. 128 (April 1, 1982), and reprinted in the *Natural Areas Journal,* 2:31–32 (1982).

Surinam

In December 1982 Richard Prum, a young ornithologist working on the social behavior of birds in Surinam, went to Bernhardsdorp at my request. He took detailed notes and photographs and spoke with some of the residents. We subsequently met to reconstruct the changes that occurred during the twenty years since my own visit. The details of the recent political events in Surinam, especially the Bouterse takeover and executions of December 1982, are based on reports from Amnesty International ("Urgent Action," December 13, 1982, January 11, 1983; *Amnesty International Report, 1983,* Amnesty International Publications, London, 1983) and "A Country of Mutes," *Time,* May 30, 1983. The sources of information used in the two accounts are at least partly independent, and Amnesty International provided a full list of the names of the victims, as well as details of exchanges it had with the Surinam government with reference to the human-rights violations. The actions of the government have been in one sense even-handed: the victims included Bram Behr, a journalist with the communist weekly *Mòkro,* as well as Bouterse's own local military commander.

Acknowledgments

FOR TECHNICAL HELP and sound advice not always taken I thank the following colleagues and friends: Freeman J. Dyson, Gerald Holton, Kathleen M. Horton, Jehoshua Kugler, William N. Lipscomb, Thomas Lovejoy, Charles J. Lumsden, Peter Marler, Marvin L. Minsky, Eviatar Nevo, Gordon H. Orians, Raymond A. Paynter, Jr., Richard Prum, Glenn Rowe, Joshua Rubenstein, Michael Ruse, Sue Savage-Rumbaugh, Lyle K. Sowls, J. Gray Sweeney, James H. Tumlinson, Barry D. Valentine, Ernest E. Williams, and Renee Wilson. This being a more personal book than my other ones, I feel it also appropriate to acknowledge with warm gratitude the staff of Harvard University Press for the effort and trust they have placed in my work during fifteen years of collaboration. Our relationship made the pleasures of authorship lasting and its pains already mostly forgotten.